GW00745987

THE MODERN DIESEL

THE MODERN DIESEL

Development and Design

edited by

D. S. D. WILLIAMS, C. Eng., M.I.Mech.E.

revised by

D. S. D. Williams, C.Eng., M.I.Mech.E.
R. J. B. Keig., M.A., C.Eng., M.I.Mech.E., M.I.Mar.E.
John M. Dickson-Simpson

LONDON
NEWNES–BUTTERWORTHS

THE BUTTERWORTH GROUP

ENGLAND
Butterworth & Co. (Publishers) Ltd.
London: 88 Kingsway, WC2B 6AB

AUSTRALIA
Butterworths Pty Ltd.
Sydney: 586 Pacific Highway, NSW 2067
Melbourne: 343 Little Collins Street, 3000
Brisbane: 240 Queen Street, 4000

CANADA
Butterworth & Co. (Canada) Ltd.
Toronto: 14 Curity Avenue, 374

NEW ZEALAND
Butterworths of New Zealand Ltd.
Wellington: 26–28 Waring Taylor Street, 1

SOUTH AFRICA
Butterworth & Co (South Africa) (Pty) Ltd.
Durban: 152–154 Gale Street

First published in 1932 by Iliffe & Sons Ltd.
13th edition 1959
14th edition published by Newnes–Butterworths,
an imprint of the Butterworth Group, 1972

ISBN 0 408 00075 9

Filmset by Filmtype Services Limited, Scarborough, England.

Printed in England by Clarke, Doble and Brendon Ltd.
Plymouth, Devon

Preface to the Fourteenth Edition

The thirteen years since the preceding edition of this work was published represents roughly one-seventh of the period during which the compression ignition engine has served mankind in an ever-growing number of applications. The class of engine to which this book is principally devoted—the automotive one—has been with us for around 40 years and is a still expanding commercial success, accounting annually in Britain for the greatest proportion of the diesel aggregate horse-power.

After this passage of time, and extending coverage of transport power requirements from the little garden tractor and the taxicab to the vehicle exceeding 40 tons (40·6 tonnes) gross weight, needing 300 bhp or more, there is no sign of a pause in development. For example there are calls for higher output from smaller bulk, including wider use of turbocharging, improved fuel economy and the ability to consume a wider variety of liquids, quieter operation, lower weight, reduced pollution of the atmosphere, and reduced maintenance requirements. There is going to be keen competition from other types of piston engine and rotary piston units in the lower power regions. At the high power end of the scale—250 bhp and upwards—there is going to be increasing competition from gas turbines especially planned for transport duties.

A feature of recent years has been the much greater employment of basically automotive diesel engines in other fields such as marine, railway traction and industrial applications. The term 'industrial' embraces mobile and transportable machinery in a big way, as well as static plant. Another demand is for prime movers for off-highway duties, which can occasion very severe conditions.

This widening pattern of demand has resulted in much larger numbers of engines being produced, in consequence of which there has been a beneficial effect on prices. Looking back over the period since the thirteenth edition of this book appeared, the price per brake horsepower paid by customers for British automotive-type diesels has risen only to a slight extent—far less than on almost any other manufactured product.

Despite the many years of diesel research and development which have marked the whole of the twentieth century there are great areas of potential improvement. In this country we have a network of research organisations which deal with both collective and individual problems, enabling us to keep in the forefront of technical progress, as is reflected in the chapters of this book. In addition many engine components from Britain are embodied in power units built in other lands. Figures published while the major portion of this volume was being printed show that the United Kingdom is the world's greatest exporter of diesel engines of all kinds. A very high proportion of the aggregate power output is in the form of automotive engines; in 1968 the UK sent overseas 42 081 automotive diesels, whilst West Germany exported 15 487 and the USA 8 049 units. Tractor diesels also were more than four times as numerous as those from the next largest producer.

At this time engineers of many nations are engaged in standardisation of weights and measures in metric terms, the state of change varying from country to country, a few having made no move so far. It is a process necessitating considerable passage of time as many engines and their components have been designed in inches, and the resulting products will be sold under those original terms for years to come. A dominant factor in this matter is the policy of the country of origin of design.

In this fourteenth edition we have given, so far as is practicable, information in both forms to facilitate matters for readers all over the world. In future the progress of metrication will result in growing numbers of engineers thinking naturally in terms of the universally standardised terms.

In concluding this preface I wish to pay tribute to my editorial predecessor, the late Donald H. Smith. What he wrote years ago as to the then future has been fulfilled in detail. He was a far-sighted practical engineer and a man of great integrity, whose character was reflected in his technical work.

That this edition may help many thousands in various parts of the world, as did the earlier ones, is the sincere hope of all of us concerned in its preparation.

Bourne End, Bucks D.S.D.W.
1972

CONTENTS

1
Development of the high-speed automotive diesel

When the possibilities of the diesel or compression ignition engine began to attract active interest for road transport purposes, as distinct from the marine, railway and industrial fields in which it was already established, some important and vital problems had to be faced. Easy starting and great flexibility over a reasonably wide speed range were obvious performance requirements, coupled with maximum fuel economy. Closely allied was the need for the lightest weight possible, together with a favourable power/weight ratio.

These requirements entailed considerable modification of existing marine and industrial engines and their equipment, and it was not until about 1931 that true road transport oil engines began to appear. Thereafter progress was continuous and more rapid in Great Britain than in any other country. The diesel quickly became almost universal in British heavy goods and passenger transport vehicles. It also became exceedingly popular in European countries but similar acceptance was somewhat delayed in the USA.

The advances in technique which led to the successful development of the road transport oil engine in turn benefited rail, marine and industrial units. In this connection references must be made to combustion chamber design, to fuel injection equipment and the solution of metallurgical problems to ensure satisfactory working life of crankshaft pins and journals, bearings, cylinder bores, and other highly stressed components. Notable improvements were made in the quality of fuel oils and in the nature of the lubricating oils required to cope with certain factors peculiar to the compression ignition engine. This technical progress was reflected in marine application which enormously increased in extent in the class of vessel requiring a compact engine with good power output in relation to the total installation weight and fuel storage capacity.

Between 1931 and 1939 the modern automotive diesel engine was developed from the experimental stage to something approaching finality in design. It may possibly be considered sweeping to use the word 'finality' in this context, but it can be justified by pointing to the fact that during the three or four years immediately prior to 1939 the number of engine types in production was generally being reduced, while several makers who had entered the market tentatively during the early development stages with more or less experimental automotive diesels had decided to withdraw or to revert exclusively to their original activities in the purely industrial fields. There was also a marked tendency towards the almost general adoption of the open combustion chamber system.

The outbreak of war in 1939 very considerably affected the automotive diesel industry. Oil engines were not so readily adaptable to military requirements as petrol engines, and their rather specialised maintenance requirements did not suit them to the immediate use of untrained and semi-skilled military personnel. Production of civilian vehicles was virtually stopped except for essential replacements. Progress was thus almost completely arrested in all branches of automotive diesel production.

In the air also, the high performance required of combat aircraft precluded further work on compression ignition engines. As a short-term possibility the

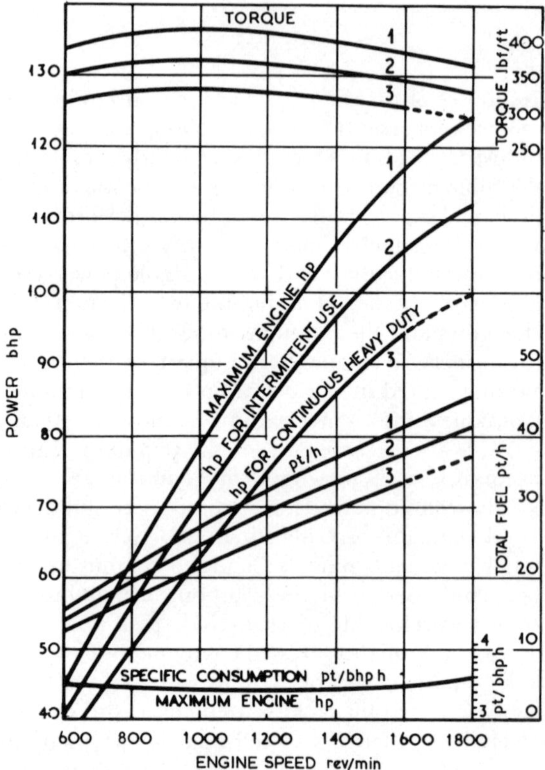

These curves show how automotive performance (1) is de-rated for heavy-duty (2) or continuous (3) marine or industrial service. They refer to a 600 in³ Leyland

aircraft diesel was not sufficiently advanced to justify wartime development in competition with petrol engines. For any desired capacity, the specific weight of the diesel was substantially higher than the 1 lb/bhp of the petrol engine.

It is true, of course, that the high thermal efficiency of the diesel engine, reflected in the low specific fuel consumption, gave it certain advantages in the matter of range. This had been illustrated by the fact that in 1938 the world long-distance seaplane record was held for a time by an aircraft fitted with a two-stroke diesel engine, although a catapult-assisted take-off was necessary. The absence of carburettors certainly eliminated some of the difficulties of high-altitude flying. Range and altitude were not so important in the early stages of the war as they became subsequently, and it was clear that improvements to existing petrol engines, to realise the advantages offered by the newly introduced high-octane fuels, provided the readiest method of achieving the performance demanded by operational requirements. In short, the diesel ceased to have any significance in aircraft development, even before the advent of the gas turbine relegated piston engines to their present secondary status.

It was in the road transport field that the diesel proved its economy and usefulness. In most countries today the heavier types of both goods and passenger vehicles are equipped with oil engines and for the most part the latest type of British heavy vehicle now has a six-cylinder direct-injection engine of from 6 to 12 l. capacity, developing from 110 to 200 bhp. It has been on this type that the success of the British automotive diesel was consolidated. The field is widening, however, because for some years now it has become exceptional to find a petrol engine in a vehicle of four tons unladen weight or thereabouts. Indeed, there are many machines of 3 tons unladen in which diesel power units are standardised. A new class of six-cylinder diesel engine of 6 to 7 l. capacity has now become established as the power unit most suited to single decker buses, its general design closely following the lines of the bigger engines. Thus the trend towards uniformity evident before 1939 was apparently hastened by the war, because most of the British engines with indirect combustion chambers were discontinued and post-war production was resumed by all the leading manufacturers on direct-injection units.

The increasing use of diesel power in medium capacity trucks has encouraged the development of a new class of engine, a four-cylinder type of from $2\frac{1}{2}$ to $3\frac{1}{2}$ l. giving a power output of from 50 to 80 bhp. In these models the direct-injection combustion chamber system is the rule although a more simple piston cavity than the toroidal form has been favoured. In this capacity class, however, the indirect combustion chamber still has a considerable following, being much in evidence in the case of units of less than 90 mm (3·54 in) bore.

In America the most commonly used diesel engine in road vehicles, particularly in buses, is a six-cylinder two-stroke with forced induction. This type has not been adopted extensively in this country and no doubt the differences in operating conditions provide the explanation. Here, as a result of our taxation system, the primary aim is to achieve fuel economy, hence the popularity of the direct-injection four-stroke. American conditions demand high speed and high vehicle performance; fuel cost is of much less pressing consequence. The two-stroke engine has a maximum power output about 66% higher than that of a four-stroke engine of approximately the same capacity. The required performance is thus obtained without appreciable increase in the installation dimensions or weight of the engine. Fuel consumption is increased not only *pro rata* to the extra power, but there is also the fact that the specific rate of consumption tends to be notably higher than that common to British engines.

Thus, power for power, fuel consumption of a two-stroke may be less favourable than that of a corresponding four-stroke. Nevertheless, two British vehicle builders have developed distinctive two-stroke units of their own design, one being of conventional in-line vertical type while the other is a horizontal, opposed-piston unit which tends to show 'Continental' affinities in its layout. Both engines have specific consumptions differing little from those associated with the best four-stroke types. It is claimed that actual vehicle consumption figures in terms of ton mile/gal set a new standard, bearing in mind that specific fuel consumption, as defined by test bed figures, does not provide a complete picture, since the final result in service is determined by extra power, reduced weight, better torque characteristics, improved acceleration and acceptance of load and smoother all-round performance. Four-stroke engines in America are based mostly on German indirect combustion chamber designs and are likewise not comparable with the average British engine in fuel economy; reduced thermal efficiency is accepted as the price to be paid for a more ready

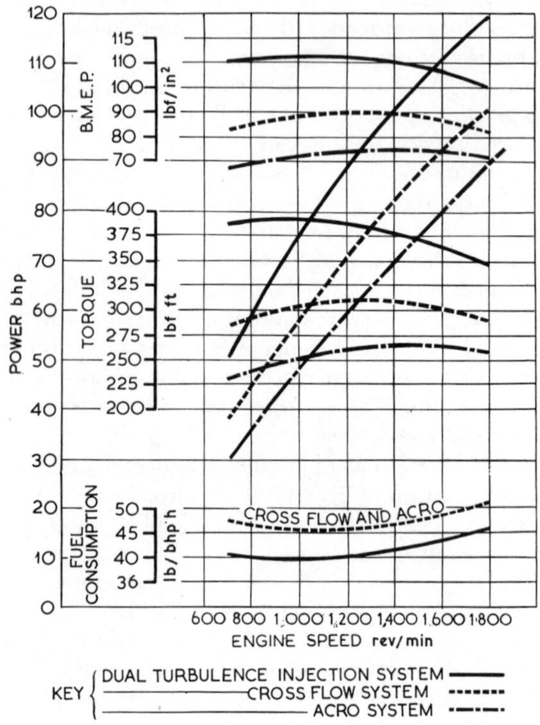

Performance improvement following development from pre-combustion to direct-injection systems in Saurer engines

acceptance of lower-grade fuels in many parts of the world to which these engines are exported.

During World War II little information could be obtained regarding the German automotive diesel industry but the indications were that development and progress were virtually arrested. Subsequent investigation after 1945 confirmed this and also revealed that most of the factories concerned had been so heavily damaged by bombing that little could be expected for some years to come. It was not until 1950 or thereabouts, indeed, that German automotive diesels began to reappear in quantity. Although one or two horizontal engines were introduced, the general mechanical layout and the previously favoured combustion systems of the antechamber or air cell types were largely retained;

indeed, it is only during the past few years that any tendency towards the adoption of the direct-injection combustion system has become evident. Certain striking innovations have been made in the development of two-stroke engines, both of the uniflow and loop-scavenged types, but it is not yet possible to assess their influence upon design in general. Likewise there are several examples of air cooling, while many of the established concerns have introduced supercharged versions of their existing engines, both mechanically driven and exhaust-turbo blowers being used, the latter predominating. It cannot yet be said that the European interest in air cooling has been reflected in British practice although supercharging is certainly receiving increasing attention.

Most of the other Continental countries had, for many years prior to 1939, based their oil-engine production on German designs and had made few original contributions to the development work that was a feature of the early 1930s. This applied particularly to Italy, France, Austria and Belgium, but not to Switzerland.

Like ourselves, the Swiss began by close adherence to German types of combustion chamber design but the influence of a successful British direct-injection engine in the early 1930s led to the development in Switzerland of the toroidal-cavity piston or 'dual-turbulence' direct-injection engine and this, as previously mentioned, has now become in one form or another the most commonly used combustion chamber design in British engines.

It would appear that the dual-turbulence type of direct-injection combustion chamber has limitations in the matter of speed because in 1937 Saurer introduced a range of light high-speed engines capable of operation at 3 000 rev/min and over, and in this design the toroidal-cavity piston with the associated induction swirl has been completely abandoned. No intake turbulence is provided at all and a simple oval piston cavity is used, the only turbulence effect being that resulting from the 'squish' of air into the cavity when the piston nears the cylinder head, where minimum clearance is allowed. Engines to this design are or have been made under licence in this country and in France and Italy.

In marine applications the conditions differ somewhat from those obtaining in road transport. The need for acceleration through a wide speed range is absent, while insistence on the most favourable power/weight ratio is not so pronounced. Fuel economy, however, is very important; likewise maximum reliability. In these latter directions the automotive-type diesel fulfils all requirements.

It is interesting to recall that one of the first British engines, of what may be called the modern type, that was applied to a road vehicle was a relatively light marine unit which, in 1929, was an innovation in its own field. Development of road vehicle engines was so rapid that in the space of a few years they were being adapted for marine application; thus the original indebtedness was repaid in full measure and when World War II came many of the high-performance engines produced by or for the road vehicle industry were re-designed for marine use and gave impressive service in coastal patrol vessels, landing craft, and suchlike applications.

Similarly the marine version of the American forced-induction two-stroke engine provided the power which enabled thousands of invasion craft to cross combat seas. The relatively small high-speed diesel has thus fully proved itself and for pleasure, fishing and utility craft its reliability, fuel economy and freedom from fire risks assure its position.

Mention must also be made of rail traction because the use of diesel power had extended rapidly prior to 1939. Not only were large high-powered diesel locomotives in use for trains, particularly in USA and on the Continent, but there was an increasing application of the smaller engines to railcars and shunting locomotives in this country. As in the marine field, there had been a similar conservatism towards development of the smaller units which fall within the scope of this book until the road transport industry entered the scene. Thereafter many of the larger road transport engines were adapted to railcar propulsion.

In this application, however, some of the performance characteristics of the road vehicle engine are not altogether necessary, while possibly an equally important aspect for rail traction is the provision of a suitable transmission of the automatic, or semi-automatic, type which will match the engine output to the torque loading on the complete installation. Obviously there is a variety of reasons why the manually controlled clutch and gear change of the road vehicle are unsuitable, one being the mechanical undesirability of breaking the torque line during gear changes, while another arises from the remoteness of the driver from the power plant. On a bus or lorry the driver manipulates clutch and gear change very largely 'by ear', but this facility becomes impossible on a railcar or diesel train where the distance between the engines and the driving position is considerable, with the result that either the exhaust note and general 'feel' are totally lost, or they are concealed by track noises.

In this connection there is still scope for the further development of electric transmission systems, but meanwhile the modern British lightweight diesel train with its multi-engined power cars is equipped with fluid coupling or torque converter drives in conjunction with automatically changed epicyclic gearing which has been developed, like the engines, in automotive practice.

There is an ever-increasing trend towards the use of diesel power units in small locomotives for work in mines and quarries, as well as for mechanical shovels, dumpers and bulldozers. Such engines are generally larger, heavier, and slower running than those fitted to road transport vehicles, but they are far more closely related to them than to the traditional types of massive industrial and stationary power units because of the wider variation in the demand put upon them, both as to load take-up and speed.

From the point of view of the absence of electrical ignition equipment, non-volatile fuel and lower exhaust temperature, the diesel engine is a much smaller fire risk than the petrol engine when working in enclosed places and in the neighbourhood of inflammable material.

Throughout the whole of the development period of the high-speed automotive-type oil engine, successive editions of *The Modern Diesel* have closely followed and recorded progress. It was the first handbook to be published setting out the basic principles in a manner which the ordinary user of petrol engines could understand. The earlier editions contained details of a great variety of experimental engines, but by 1939 the industry was tending towards concentration both as to types of engines and as to the number of manufacturers in active production. Nevertheless, throughout the following pages particulars are retained of certain of the development types that have had more than a passing influence on the history of the subject. A study of these will enable the reader to arrive at a better understanding of the principles involved.

As to the future, it is not easy to visualise any changes from the current types of a magnitude comparable with those which marked the first 10 years of development. It may well be that forced induction will become more evident, not only on two-stroke engines, on which it is essential, but on four-stroke units as well. Mechanically driven superchargers have proved their effectiveness and reliability in both two-stroke and four-stroke applications on land and at sea. Power output can be increased up to 50%, or even more if conditions allow the principle of heat exchange to be applied to effect cooling of the charge air between the supercharger and the cylinders. One of the attractive possibilities of forced induction is so to plan the equipment that the greatest boost is effected at the low-speed end of the engine's speed range; this is most beneficial in road and rail services of diesel engines. With this scheme much gear changing is cut out. Generally speaking, the cost of supercharging—both prime cost and extra maintenance—has to be weighed up to appraise its worth on any particular engine. In some cases supercharging offsets atmospheric pressure reduction at altitudes and therefore has appeal for work in mountainous territory.

A growing development in forced induction is the turbo-blower system in which a single-stage turbine is driven by the exhaust gas. Directly coupled to the turbine is a bladed disc-type blowing element connected by a duct to the air intake manifold. Exhaust energy that would otherwise be wasted thus provides 'free' blowing power and there is a resultant improvement in overall thermal efficiency, maximum power output being greatly increased at high load and speed ratings. Indeed, the turbo-blown diesel is at its best under heavy load at constant high speed, which no doubt accounts for the fact that turbo-blowing is more commonly used on heavy duty railcar and industrial installations than it is on otherwise identical power units used under the fluctuating load and speed demands of normal road transport vehicle operation.

At the other end of the range rather less spectacular progress has been made with small engines, for although there have been one or two recent introductions in the four-cylinder 1·7 l. class, in general it would appear that individual cylinder capacity cannot be reduced much below 0·4 l., so that for six-cylinder engines of automotive type a total swept volume of about 2·5 l. is the smallest practicable size. It is true that certain vehicles have been equipped with 'small' engines of this type by reducing the number of cylinders of this size to provide a 2·25 l. three-cylinder or a 3 l. four-cylinder unit. In marine applications there are single- and twin-cylinder engines, also of about the same cylinder volume; all these engines are highly successful within their particular spheres, but the true small engine is rare. It may be mentioned that the three-cylinder diesel in this size range has of late become a very important factor in the development not only of the 1 ton delivery van but even more so in connection with small agricultural tractors.

Almost all the established British makers of the larger type of automotive diesel have introduced smaller units since 1948, mainly in the 5 l. class. Practically without exception these have been six-cylinder units, although one prominent maker decided on a four-cylinder type on the grounds that a six in this capacity class is too small in the bore for best results with the direct-injection system and too restricted as to crankshaft bearing spacing to provide optimum life. This view appears to have been borne out by a later tendency to increase the size of these medium-capacity engines rather than to reduce them still further. By implication, therefore, the likelihood of the appearance of really small diesels of under 2 l. in any quantity, and suitable for private cars or light vans, is doubtful; the possibility cannot be dismissed out of hand, however, in view of certain established Continental examples, although one of the earlier German models has recently been withdrawn. There is, nevertheless, a growing British interest in four-cylinder diesels of just over 2 l. capacity. All the small engines so far made have had antechamber or air cell combustion systems.

2
The term 'diesel'

Before approaching the subject of this book in detail it is desirable to define the term 'diesel engine'. Unfortunately the name is surrounded by a mass of acrimonious controversy which is probably unequalled in the whole field of mechanical development. At every stage there have been rival claims to the invention of this or that aspect of what we now call the diesel cycle, and even in quite recent times there have been rumours of litigation concerning some of the quite modern developments of the basic system.

Of the earlier controversies, it can fairly be pointed out that the tragic end of Dr. Diesel absolves him from any share in the disputes, for on the night of 29–30 September 1913 he disappeared from the Antwerp to Harwich steamer in circumstances which have never been explained.

It would seem that from time to time various subsequent enquirers were anxious to publicise their own discoveries by revealing what they regarded as prior claims to certain aspects of the diesel engine as we now know it and which was based originally on the hypothesis first published by Dr. Rudolf Diesel under the title 'Theory and Construction of an Economical Thermal Motor'.

The first contentious point that arose was the insistent claim put forward that Diesel's subject matter was not original but in effect a transcript of theories propounded by Professor Linde of Munich University. Diesel was born in Paris of German parents in 1858 and studied at Augsburg and Munich, after which he went into the engineering works of Sulzer Bros. in Winterthur, Switzerland. He developed the theories of 'the economical thermal motor' about 1890 and took out various patents, including a British patent in 1892, the essential part of which reads:

> Motive work by means of heated air . . . compressed to so high a degree, that by the expansion subsequent to the combustion the air is cooled to about atmospheric temperature, and that into this quantity of air, after its compression, fuel is gradually introduced. . . . At this compression the temperature becomes so high that the fuel employed is spontaneously ignited when it comes into contact with the compressed air.

It is clear from this patent that Diesel's claims were for an engine designed to conform to a particular thermodynamic hypothesis and his chief purpose was to avoid the two main sources of heat loss in an internal combustion engine, controlling maximum temperature by introducing the fuel gradually and discarding only cooled exhaust gases. He argued that a large increase in thermal efficiency was impossible if any fuel–air mixture was present before ignition, owing to the amount of excess air allowable being limited by the possibility of igniting the mixture, compression and expansion ratio thus being limited entirely by the danger of preignition.

The basic requirement of the diesel cycle was that at maximum compression pressure the fuel should be admitted in such a way that combustion would be maintained at *constant pressure* in the cylinder during the burning period, whereas in all previous internal combustion engines the fuel had been burnt instantaneously (or nearly so) without change of volume, that is, with great pressure rise at *constant volume*. Diesel's cycle was not primarily conceived with the idea of securing self-ignition of the fuel by the method of using a very high compression ratio; the self-ignition was incidental to the high compression ratio necessary to comply with the thermodynamic principle which he propounded and the aim of the cycle was to maintain the maximum compression pressure by introducing fuel over the most effective angular travel of the crankshaft during the power stroke.

In theory the pressure rise during combustion should not have greatly exceeded the maximum compression pressure and the result should have been a smooth development of power more akin to the thrust of the steam engine than the explosive blow on the piston of the normal internal combustion engine.

Diesel's theory received wide and enthusiastic acceptance among engineers and experimental work was undertaken both by Krupp and at the M.A.N. works at Augsburg. To quote his own words: 'In 1897, after four years of difficult experimental work, I completed the first motor in the Augsburg works.' The engine was first publicly exhibited at the Munich Exhibition in

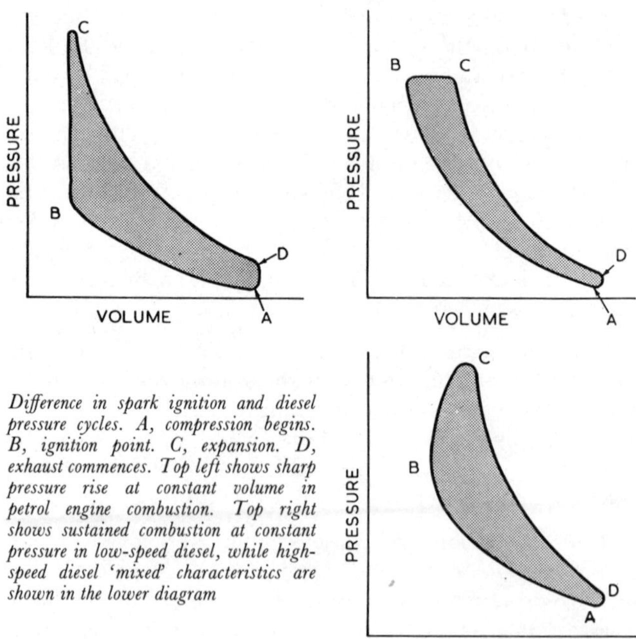

Difference in spark ignition and diesel pressure cycles. A, compression begins. B, ignition point. C, expansion. D, exhaust commences. Top left shows sharp pressure rise at constant volume in petrol engine combustion. Top right shows sustained combustion at constant pressure in low-speed diesel, while high-speed diesel 'mixed' characteristics are shown in the lower diagram

1898 and a year later Diesel established his own works at Augsburg. Nevertheless, it is true to say that *no automotive engine strictly supporting the Diesel theory has ever been built*!

As originally laid down the diesel cycle provided for combustion at *constant pressure*, whereas the ordinary petrol engine gives combustion at *constant volume*, its combustion taking place rapidly when the compressed air–petrol mixture is ignited by an electric spark. These two conditions of combustion are shown graphically by the accompanying indicator diagrams.

In petrol engines, with ignition timed at the usual angle before top dead centre, combustion should be very nearly at constant volume, as shown by the vertical part of the left-hand diagram. Only in low-speed diesel engines can the combustion process be approximately at constant pressure, as shown by the horizontal part of the right-hand diagram.

Modern high-speed compression ignition engines, however, come between these two conditions according to the instant at which fuel injection is commenced. If injection is commenced just before dead centre the curve obtained is flat at the top, but as the moment of commencement of fuel injection is advanced, as it has to be in high-speed engines, the curve can be made sharper and sharper, with correspondingly higher maximum pressures, until practically constant volume combustion is attained, as is shown in the third diagram.

From time to time there have been attempts to show that Diesel was anticipated, notably by Akroyd-Stuart, an English engineer, born in 1864, who took out various patents, one of which, dated 1890, described

> . . . means for preventing the premature or pre-ignition of an explosive charge of combustible vapour or gas and air when a permanent igniter (such as a continuous spark or a highly heated igniting chamber) is in communication with the interior of the cylinder, by first of all compressing the necessary quantity of air for the charge, and then introducing into this quantity of compressed air the necessary supply of combustible liquid vapour or gas, to produce the explosive mixture.

The difference here appears to be that whereas Diesel worked from a strictly logical thermodynamic hypothesis and propounded an engine from theoretical principles, Akroyd-Stuart simply aimed to eliminate a known defect of existing engines without apparently recognizing the basic principles involved. Diesel realised that self-ignition was fundamental to the compression pressure that had to be attained to secure the desired result, whereas Akroyd-Stuart was endeavouring to avoid self-ignition even in low-compression engines and did indeed rely upon an external ignition source, or a 'hot bulb'. The gain in thermal efficiency resulting from high compression ratio was not his immediate aim.

Notwithstanding that a high-speed engine based upon the Diesel theory has never been fully realised in practice, there is no reason to quarrel with the general adoption of the name 'diesel' as a generic term for all engines in which pure intake air is compressed to a pressure sufficient to achieve self-ignition of an injected, non-volatile fuel. For a time there was a tendency to use the term 'oil engine' but as this was not fully expressive in regard to the special type of compression ignition system involved, the use of 'diesel' has again become common.

3
General and historical survey

The rapid development of the automotive diesel engine was one of the most notable features of engineering progress during the 1930s. Although the particular type on which so much thought has been expended is generally termed 'the high-speed diesel', the description 'high-speed' is only comparative; for the most part the class of engine under review has a practical range of speed up to about 2 500 rev/min.

This speed may appear rather low to those who are only familiar with petrol engines, for even the workaday types of spark ignition units usually run up to over 4 000 rev/min, while over 6 000 rev/min is by no means unusual in recently developed car and motor-cycle units. The high-speed oil engine as now used is essentially a recent development, and it owes its growth from a mechanical point of view to two important factors, the commercial production of the small high-pressure metering (or measuring) fuel pump and injection nozzles, and the vast store of mechanical and metallurgical knowledge already embodied in the high-efficiency internal combustion engine.

Of these two important stepping stones undoubtedly the fuel pump is the key to the whole forward movement. As is fully explained in later chapters, diesel engines have the charge of oil for each power stroke delivered to the cylinder under high pressure at the required moment, and the only controls of speed and power are the regulation of the quantity of oil at each injection, and of the moment of commencement of injection. If a single-cylinder engine of about 101 mm (4 in) cylinder bore runs at 2 000 rev/min, a spot of oil less than the size of a grain of rice' must be injected at an exact moment in the cycle of operations 1 000 times a minutes, while when the engine is ticking over at 300 rev/min the size of each charge injected will be smaller than the head of a very small pin. Between these limits the pump delivery must be capable of infinite variations and when a multi-cylinder engine is involved each cylinder must have an exactly equal charge for any setting.

Although it is difficult to trace the originator of the jerk pump, the invention has been attributed to a British engineer, the late A. E. L. Chorlton; claims have also been made that Richard Hornsby & Sons, of Grantham, were using jerk pumps in 1891 and that Ruston, Proctor & Co. used a similar type of pump in 1909. However, it is to Robert Bosch that the credit must go for first making such a pump commercially. Indeed, he made high-precision injection equipment available even before there was a demand for it.

Controversy of this nature appears to have followed upon every development associated with the diesel engine, and as recently as 1945, in the course of investigations by the British occupying authorities into the diesel engine industry in Germany during the war, a new claimant appeared in Franz Lang, inventor of the Acro and later the Lanova combustion chambers. Among other things, Lang stated that it was he who made the prototype fuel injection equipment which he introduced to Robert Bosch prior to 1912.

Given the metering pump, the mechanical aspects of the compression ignition engine presented no insuperable problems to those conversant with the design of the high-speed petrol engine, and so the way was clear. Why was it taken up and developed?

The spark ignition volatile-fuel engine, excellent as it is, is wasteful. Fuel has a certain heat value, and heat is work. The petrol engine delivers only 22–25% of the theoretical work value of its fuel, while the oil-using c.i. engine delivers 30–36%. Thus the latter is said to have the higher thermal efficiency, which in terms of ordinary usage means that there will be a more economical consumption figure for a given load.

The comparatively low temperature at which petrol ignites limits the compression ratio in the petrol engine to a comparatively low figure, above which pre-ignition occurs. But it is known that the higher the compression the more efficient an engine becomes, and this gives the diesel engine its second advantage, because only pure air is compressed so that there is nothing to pre-ignite, the oil injection being timed to take place when it is wanted and not before.

The torque of the diesel engine is high, torque being a capacity for turning, or good pulling power. The oil engine pulls well over all its speed range; in particular, it is not easily stalled when the speed drops, which is a fault with the petrol engine. Consequently we can begin to appreciate its attractions. It is economical with fuel, it is a good puller, and a robust and mechanically accurate metering pump replaces the carburettor and electrical ignition apparatus.

The fact that it consumes a fuel which is not dangerously inflammable makes it additionally attractive in many ways, particularly for small pleasure and utility craft on the water, where fire has such dire consequences. That the fuel is not volatile is also important in the tropics, where diesel-engined lorries and buses can operate over great distances with safety because there is little evaporation loss from the tanks, a loss which may account for 50% on one journey when petrol is used. Increased range per tankful is an important factor,

THERMAL ADVANTAGE OF THE DIESEL CYCLE

	Petrol Engine	*Oil Engine*
Heat Profit	%	%
External work done	23	34
Heat Loss		
Internal work and friction	10	11
Cooling water and radiation	34	31
Exhaust gas	33	24
	100	100

Showing how heat (fuel) put into the engine is returned. The external work delivered represents an economy of 48% in favour of the oil engine

especially in undeveloped countries. Diesel exhaust is also far less noxious than that of petrol engines.

Most of the credit for the pioneer work in the adaptation of the diesel engine to road vehicles must be given to Germany. Rudolf Diesel had always maintained that his system could advantageously be applied to road vehicles. Indeed, in 1913, only a short time before his death, he wrote: 'I am still firmly convinced that an engine will be developed for road vehicles and when that is done I shall consider the aims of my life have been realised.' Some of his earliest work had been directed to road vehicle usage and an engine for this purpose had been made in 1898; he was also directly responsible for another built in Switzerland in 1910. These efforts were not, however, successful.

Another road vehicle engine was meanwhile being prepared by the M.A.N. organisation in Germany, and it was demonstrated at the Turin Exhibition of 1911. Like the others it was not acceptable to practical engineers, mainly because of the complication of its air blast injection system. These objections convinced Diesel and his collaborators that air blast injection must be discarded on small engines and that his earlier schemes for the 'solid injection' of liquid fuel would now have to be more fully explored. About this time Bosch had his prototype injection pump in being and the stage was set for a great forward step.

However, the 1914–18 war intervened and further development of road transport diesel engines ceased, the German automobile factories being diverted to the production of arms, aircraft engines and submarine power plant. Submarine engines, of course, were diesels and familiarity with the type was thus considerably widened, this being subsequently reflected in the appearance of several makes of engine specifically designed for road vehicles. Notable among these were a 50 bhp four-cylinder Benz and a 40 bhp two-cylinder M.A.N., both of which appeared in 1922 and were characterised by their crude and untidy appearance.

The 1921 Benz 50 bhp four-cylinder diesel engine

The M.A.N. engine, however, was quickly cleaned up and developed into a very modern-looking four-cylinder unit with integral clutch and gearbox which appeared at the Berlin Show in December 1924. By contemporary standards it was a high-speed engine, although its 1 050 rev/min would not warrant that classification today. In appearance, size, and detail design it was undoubtedly the first definitely automotive diesel to be produced. It developed 45 bhp from about 5 l. swept volume and was very little heavier than the equivalent petrol engine. Consumption was stated to be 181 g/bhp h

In both appearance and combustion system this direct-injection M.A.N. anticipated the modern automotive diesel. It was exhibited in 1924

(0·4 lb/bhp h) and direct injection was used. Because of its low specific consumption the Bavarian Post Office placed a contract for the supply of engines for their post buses.

Having regard to the shortage of petrol in Germany after World War I, other large operators showed the greatest interest in the M.A.N. engine and by 1932 the firm completely discontinued the production of petrol-engined heavy vehicles. Meanwhile the Mercedes and Benz concerns had amalgamated and had produced improved four- and six-cylinder units with the antechamber combustion system, the prototype 50 bhp engine already mentioned having provided convincing demonstrations, in bus and lorry chassis, in long runs from Berlin to Stuttgart and Munich. By 1928 the Mercedes-Benz was well established and had already attracted considerable attention in this country.

Several other road transport diesels were produced—mainly in Germany—but, having regard to its subsequent influence upon automotive diesel history, particularly in Britain, the most important was the Saurer from Arbon in Switzerland. The Saurer concern had intimate contacts with M.A.N. of Nuremberg because that firm had adopted the well tried and successful Saurer chassis designs in which to install its engines when it entered the heavy vehicle market. Saurer also had close relations with Robert Bosch and it was natural they should actively seek to develop the diesel. Their first effort was to convert a Saurer petrol engine in 1923; they then produced their own design, which embodied the Acro combustion chamber then being sponsored by Bosch.

In other parts of Europe the economic incentive to explore diesel possibilities was less pressing than it was in Germany. No really important contribution to the movement was made in France, Italy, Austria, Belgium or Holland prior to 1930, and in general these countries have mainly applied themselves since that date to the production of diesel engines under German, Swiss or British patent licences. Some striking developments in two-stroke engines as well as in fan-cooled four-stroke units have appeared in recent years from Austria and Czechoslovakia. Little original contribution to diesel progress, however, has been made by the USSR and so far as is known the engines being produced there closely approximate to German and American types of well-known design. In America the availability of great quantities of cheap petrol resulted in even less incentive towards the diesel system for automotive engines and it is all the more remarkable, therefore, that long before the American automobile industry became diesel-conscious, one of the most outstanding small marine diesels was produced in the USA, embodying all the features which characterise the automotive type. This engine, the Cummins, now has a prominent position in the automotive field, and is made in several countries other than the USA.

Serious interest even in Great Britain did not manifest itself until 1927 or 1928, when the combination of successful diesel development in Germany and increasing economic pressure at home directed thought towards the subject of fuel conservation. Even then only a few far-seeing individuals accurately assessed the possibilities of the diesel for automotive purposes. It is true that a certain amount of work had been done in the light marine field, but the conditions in that industry were not conducive to rapid development. This was because marine motors run at a fairly constant load and speed; high revolutions are not called for, nor are there overwhelming advantages in extreme lightness of construction. Engine demand in the marine field is relatively small.

The interest of a strong, wealthy, and highly progressive industry was needed and, happily, some of the designers of heavy transport vehicles realised the possibilities of an engine that would run a 12 ton load for 12 miles on a gallon of fuel, against a petrol vehicle's 5 miles/gallon. British makers of heavy chassis were at the height of technical success in petrol-engined transport vehicles, while several well-established British makers of marine and stationary diesel engines had for some years devoted attention to adapting them to purposes for which higher speeds and greater flexibility were required. Among these efforts may be instanced the application of Beardmore high-speed units to railcars and airships.

In 1928, the Associated Equipment Co. Ltd., of Southall, built a six-cylinder c.i. engine of their own design and ran it experimentally in a bus from December of that year, conveying employees to and from their works. The engine, which was built under Acro licence, was later fitted to a heavy lorry and further tested.

A Mercedes-Benz diesel lorry (with trailer) earned the Dewar Challenge Trophy for 1928 as a result of an RAC test. Over a distance of 691 miles (1 132 km) it consumed fuel at the rate of 13·48 mile/gal (4·77 km/l.), equivalent to 156·5 gross ton mile/gal (56·27 tonne km/l.).

At the Shipping, Engineering and Machinery Exhibition held in September 1929, the first British examples of the modern diesel engine appeared in self-contained form for sale to the public. There were three four-cylinder units, a 36 bhp Gardner direct-injection cold-starting engine, an 80 bhp Gleniffer four-cylinder which had an ingenious air-starting motor and a 40 bhp Ailsa Craig with electric starter. There was also a six-cylinder American Cummins of

60 bhp which, after modification, was applied to a racing car that exceeded 100 mile/h at Daytona. The Gardner engine was installed in a bus by T. H. Barton, of Nottingham, and it did good work.

At the Public Works and Transport Exhibition, also held in 1929, the Kerr Stuart heavy lorry, powered by a Benz-type engine made by McLaren's, of Leeds, was exhibited, while early in the following year the Sheffield Corporation put into service a Karrier six-wheeled bus chassis with a Mercedes-Benz engine. The late W. H. Goddard, A.M.I.Mech.E., of Leeds, an ardent pioneer of the heavy-oil vehicle, was responsible for the conversion, which gave good results.

In quick succession Leeds and Manchester Corporations ordered Crossley double-decker buses fitted with Gardner six-cylinder engines, and obtained 10–12 mile/gal (3·6–4·2 km/l.) against half the distance of petrol buses on town work, while Frank Dutson of Leeds commenced the fitting of Gardner

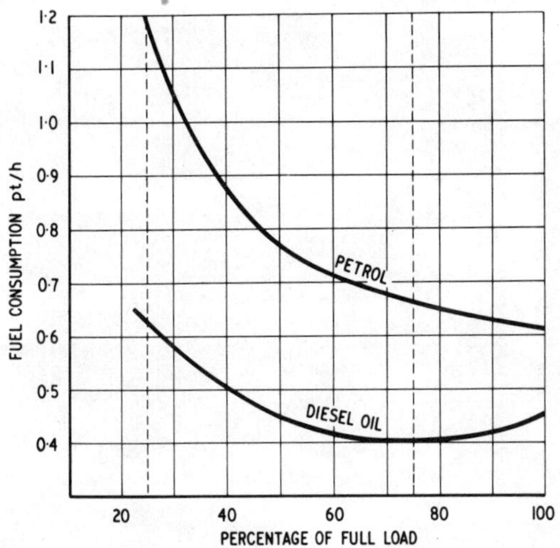

Curves showing comparative fuel consumption of equivalent 130 bhp six-cylinder AEC engines running on petrol and diesel oil (circa 1932)

engines to existing lorries as a business undertaking and demonstrated remarkable economies.

Thus from this rapid sequence of events grew an essentially British movement that set the entire heavy-vehicle industry by the ears. Factory after factory hastened into experimental work, so that by the end of 1931 over 40 makers in Great Britain, Europe and America were producing diesel engines for road transport work, and most manufacturers of buses and lorries offered them as alternatives to petrol engines. Perfection was not reached at the first attempt, and, naturally, the pioneer vehicles mentioned had their limitations. Marine-type engines were slow, heavy and bulky, but in a decade of development weights were reduced from about 10·8 to 4·5 kg (24 to 10 lb) or less/bhp, which was comparable with the best petrol practice in the case of road transport vehicle engines.

Many makers responsible for this intensive development of automotive diesel engines had no experience with their marine application, but it was not long before marine engineers were turning to the road transport field for the benefit of the research work that had been carried out so extensively therein. In the course of its development the automotive-type engine has become the accepted form of power unit for the average class of light craft, both for pleasure cruising and serious work.

Another direction in which the automotive class of diesel engine has revolutionised the scene is in industrial applications employing units rated at up to 200 bhp for stationary, mobile and self-propelled machinery. In the 'off-highway' plant category fast running diesels of up to 500 bhp are employed, and the general trend is for higher power/weight ratios. Tractors are almost wholly diesel powered today.

By virtue of its quantity production and the world-wide availability of spares and service, the automotive engine offers a self-contained and easily

First all-British compression-ignition lorry—a W. D.-type Leyland converted with a 50 bhp Gardner light marine engine by Frank Dutson, of Leeds, in 1930

maintained power plant at low first cost and minimum operating expense. Most well-known automotive engines are available either mounted on bed-plates or assembled in enclosed 'power-pack' housings which enable them to be operated under almost any conditions so long as there is a firm and level site to stand them on. With direct-coupled generators they provide a relatively inexpensive standby outfit for emergency electric lighting of large buildings and works or serve as portable power plants where energy is required temporarily over scattered sites.

Railway uses of automotive type engines are somewhat analogous to those in road vehicles. It is true that large, low-speed diesel engines were applied to railway working even before the automotive diesel came into being, but there was no realisation, apparently, of the possibilities of a specially developed light, small, but high-powered engine for railcar work and shunting locomotives. When the automotive types were forthcoming they were applied with success to these uses and indeed may be said to have opened up certain hitherto unexplored possibilities. Particularly has the diesel-powered development of the current type of multi-unit train been made possible by the availability of automotive engines. A prime example is the large-scale adoption by British Rail; that organisation's railcar and multiple-unit fleet embodies over 4 000 cars powered by diesels of fundamentally road transport types.

For private cars the use of the diesel engine has been very restricted. Among the chief aims of the motor-car designer have been speed, acceleration, silence, and smoothness of operation, and these qualities cannot lightly be exchanged for a fuel economy that becomes of increasingly less importance as the weight of the vehicle decreases.

In the matter of silence and smoothness of running, particularly when idling or moving very slowly, the petrol engine has inherent advantages, owing to the principle it operates upon. The carburettor-throttle control of the petrol engine is a form of variable compression control, in that at small openings it so restricts the flow of air through the intake that only a small quantity is admitted to the cylinder, with the result that the compression pressure is relatively low and the running of the engine, when idling, for instance, is 'soft' and unobtrusive. Exactly the opposite condition applies to the diesel, for its air intake is always wide open and as there is more time for the air to enter the cylinder at low rev/min the compression pressure is actually higher under these conditions than at any other time, with the result that the engine tends to run more roughly the lower its speed. Other things, such as fuel/air ratios under idling conditions, also favour the petrol engine in the matter of unobtrusive running.

Nevertheless, several cars were run experimentally both in Europe and in America during the development stages of the diesel automotive engine and the first genuinely usable car of this type operated in England was a Bentley sports saloon powered by a Gardner 4LW engine which competed successfully in the 1932 Torquay Rally and also in the 1933 Monte Carlo Rally.

Other cars were converted and a lightened AEC bus engine in 1937 was used in a car which, with Eyston and Denly driving, averaged 105·59 mile/h (169·7 km/h) for 1 h and 97·05 mile/h (156·1 km/h) for 24 h at Montlhéry. Prior to this a four-cylinder Perkins diesel in a racing car had attained a speed of 94·76 mile/h at Brooklands in 1935. This make of engine was also used experimentally in several private cars with good results, while in 1936 a 2·8 l. four-cylinder Mercedes-Benz diesel-engined car was exhibited at the Berlin Show. Another car, the Hanomag, also appeared in Germany, while a Citroën was introduced in France the following year. Within the past 15 years diesel-engined small cars of 'family' size were in production in Germany by Mercedes-Benz and Borgward and there was also the Fiat, of similar type, in Italy. All three had four-cylinder engines of about 1·9 l. The Borgward was withdrawn after a few years' work and little is now heard of the Fiat. In Great Britain a considerable proportion of the taxicabs in London, Manchester and other cities has been initially equipped with four-cylinder engines of about 2 l. capacity; many such British engines power taxis in Belgium and other European countries, including the USSR. Taxi service is exceedingly favourable to diesel performance characteristics and fuel consumption has improved from about 17 to nearly 30 mile/gal (27·3 to 48·2 km on 4·5 l.) of fuel.

So far no mention has been made of objections to the diesel engine. The early examples were heavy, slow, subject to severe vibration, lacked acceleration, fouled their injection systems and were subject to a phenomenon known as diesel knock. Most of these faults have been eliminated. Starting, too, especially in the case of engines with the more complicated types of combustion chambers, was rather difficult. On the other hand the direct-injection type of engine, now almost universal, is probably easier to start than a petrol engine of equivalent size.

Criticism was also directed against smoke in the exhaust, but here again knowledge of how to secure more complete combustion has increased to the extent that a well-designed engine in good condition will emit no visible exhaust; moreover, it could always be said that the fumes, if obvious, were not dangerous, being practically free from the invisible but poisonous carbon monoxide in the exhaust gas of petrol engines. British engines have provided a lead in complete combustion and freedom from exhaust smoke. In this connection the realisation of the importance of turbulence, notably by Ricardo, cannot be over-emphasised.

Finally, a word on the economy of the diesel engine is desirable. Some early propagandists tended to overstress claims based on the work/fuel-cost ratio. This form of comparison was altogether unfair to the petrol engine when petrol in Britain was taxed at 8d per gallon, while oil fuel was duty-free and subsequently taxed at 1d per gallon. For some time it had been expected that if heavy oil came into wide use for purposes normally served by petrol engines, its higher taxation would come under consideration. These expectations were fulfilled in 1935, and though previously the cost figure was a particularly advantageous argument, it was always adventitious in character. The only sound fuel-economy comparison between the diesel and the petrol engine is the work/consumption ratio—on that basis the diesel has an advantage of 40–60%, which is a true reflection of its thermal efficiency and favourable operating characteristics.

It is notable that the increasing number of diesel engines in city traffic coincided with a considerable agitation on the part of certain organs of public opinion regarding the possible adverse effect of diesel fumes on health, particularly in connection with a disturbing rise in the incidence of lung cancer which had lately become evident. Investigation by responsible medical scientists produced negative results; precise knowledge on the toxicity of diesel exhausts is available in the British official studies of and regulations concerning diesel engines.

4
The compression ignition cycle

Anyone familiar with the working of the ordinary petrol engine is aware that the motive power is produced by igniting above the piston in each cylinder a mixture of petrol vapour and air, the resultant combustion of which causes an expansion of gases exerting pressure on the top of the piston. By this means the heat released by the fuel is transformed into work, the reciprocating movement of the pistons being converted into rotary movement of a crankshaft by means of the connecting rods.

The four-stroke cycle is the operating sequence more commonly employed. It requires that each cylinder be provided with an inlet valve to admit a combustible mixture of petrol and air drawn by the suction of the engine from a mixing device known as the carburettor into the space above the piston, while there must also be an exhaust valve to control the outlet through which the residual products of combustion are ejected after they have done their work.

To describe the complete cycle as it occurs in any one cylinder it may be assumed that the piston is at the top of the stroke and that both the inlet and the exhaust valves are closed. As the piston descends the inlet valve opens so that a charge of 'mixture' can flow into the cylinder. This is called the suction (or induction) stroke, the charge flowing under the influence of external atmospheric pressure. Immediately after the piston has reached its lowest position and begins to rise again the inlet valve is closed so that the mixture can be compressed by the rising of the piston on the compression stroke. As the piston once again approaches its topmost position an electric spark is caused to pass between the points of a sparking plug located in the top of the cylinder in a zone called the combustion chamber. The spark ignites the mixture and as both valves are closed during its combustion, there is a considerable pressure rise. As a result, the piston is forced back again down the cylinder, this movement being termed the expansion or power stroke.

Before the piston reaches its lowest point, the exhaust valve is opened, pressure is released, and as the piston again rises it sweeps the residual gas through the exhaust outlet, whence it is led through a silencer to the atmosphere; this is the exhaust stroke which completes the continuously repeated cycle. Thus it will be seen that there is one power stroke for every four strokes of the piston or two revolutions of the crankshaft.

It should be noted that the size of the combustion chamber above the piston (known as the clearance space) is considerably smaller than that part of the cylinder in which the piston travels. Thus at the end of the compression stroke the explosive mixture is compressed to a pressure varying from about 5·5 to 7 kg/cm^2 (80 to 100 lbf/in^2). By such compression the mixture is heated to some extent, resulting in better vaporisation of the fuel, and more intimate mixing of the vapour and air; it also enables the engine to work with higher thermal efficiency, to which further reference is made later.

Incidentally, the ratio of the total volume of the cylinder above the piston when it is in its lowest position to that of the combustion or clearance space when it is at its highest position is known as the compression ratio, and in modern commercial vehicle petrol engines varies from about 6:1 to 8:1.

The operating cycle of the four-stroke diesel is identical in so far as the sequence of strokes is concerned but there is a major difference in principle. On the induction stroke only pure air is drawn into the cylinder through the inlet valve port. Then, because the clearance space is so much smaller than that in a corresponding petrol engine, the compression ratio is considerably higher, being not less than 12 : 1 and even in some engines as high as 21 : 1. The final pressure of the air charge at the end of the compression stroke may

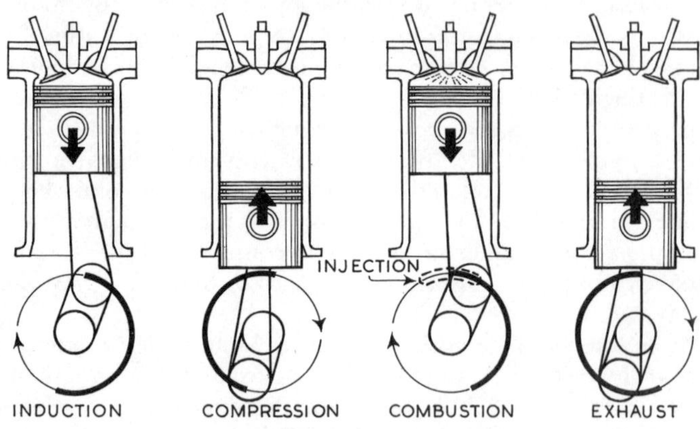

Four-stroke cycle of operations in a diesel engine during two revolutions of the crankshaft

be 35 kgf/cm^2 (500 lbf/in^2) or more, with a corresponding temperature rise to not less than 1 000° F (538° C), which is sufficient to ignite finely atomised but non-volatile fuel oil without the assistance of an electric spark.

Just before the piston reaches the end of the compression stroke the fuel oil is injected into the combustion chamber through an injection nozzle, or sprayer, mounted in the cylinder head in much the same position as the sparking plug in a petrol engine. The pressure at which the fuel is injected must necessarily, of course, be higher than the combustion pressure, usually not less than 140 kgf/cm^2 (2 000 lbf/in^2). During injection the fuel is split up into finely divided particles, and the mixture of these with the air forms an explosive charge that is ignited by the heat of compression.

Injection is continued for a short period, during which the piston passes its highest position and begins to descend on the power stroke. The expansion of combustion products only begins to have effect when the piston has passed the top of its stroke, and during the injection period the burning fuel maintains a more or less constant pressure somewhat above that of compression; when the fuel is cut off expansion of the gases continues, the pressure falling as the piston descends. The overall effect on the piston is a more sustained pressure than that associated with the combustion process in a petrol engine. The effective part of the power stroke is concluded before the piston descends to its lowest point, and the cycle is completed when the exhausted gases are expelled by the rising piston.

To inject the fuel a special type of pump driven by the engine is employed, and this is a distinguishing feature of the diesel engine, just as the carburettor

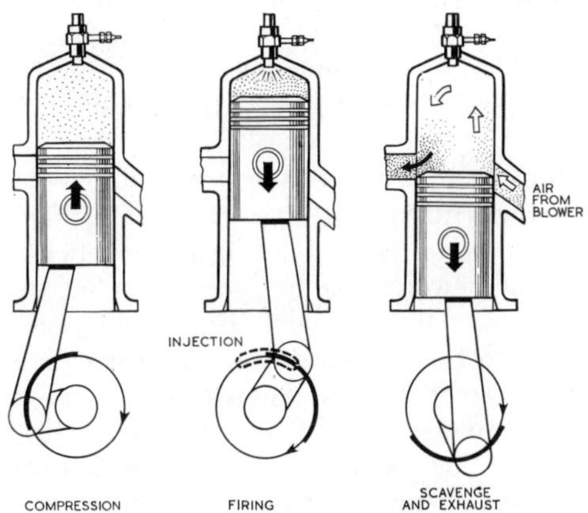

Cycle of operation in a loop scavenge two-stroke diesel engine

and the magneto or coil and distributor ignition apparatus distinguish the normal petrol engine.

Mention must also be made of the two-stroke cycle, for it is particularly adaptable to the diesel system. The two-stroke cycle is so called because the functions of the normal four strokes are effected in two piston strokes, an up-stroke for intake and compression, and a down-stroke for expansion and exhaust.

In the simplest form of spark ignition two-stroke petrol engine the underside of the piston and the crankcase are used as a pump, the rising piston drawing mixture into the crankcase through a cylinder port uncovered by its skirt while at the same time compressing a previous charge in the cylinder. Combustion and expansion drive the piston down again on the power stroke, ports in the cylinder wall being uncovered as the piston nears the bottom of the stroke so that the residual gas can escape under its own pressure to the silencer. At the same time a transfer port is opened through which the now compressed mixture in the crankcase can flow into the cylinder above the piston, providing both the new charge and assisting to expel or scavenge the remaining spent gases.

Engines of this type are familiar as the power units of lightweight motor cycles and small portable blowers, pumps or other equipment. The system makes for a very simple engine having no valve mechanism, since all intake and discharge functions are controlled by the co-operation of the piston in relationship to ports in the cylinder wall. This 'three-port' two-stroke arrangement is not thermally efficient, however, being subject to serious losses arising from incomplete scavenging of the exhaust and also to actual loss of fuel 'mixture' by mingling of the new charge with the outgoing exhaust gases.

In the case of the two-stroke diesel, however, only pure air is dealt with and by delivering excess air from a mechanical blower to the intake, effective scavenging is possible without loss of fuel, although at the cost of whatever power is absorbed by the blower. Crankcase compression is not used in automotive-type two-stroke diesels; instead, the intake air, under slight pressure, is forced directly into the cylinder through ports uncovered by the piston near the bottom of its stroke. If exhaust is released by way of piston-controlled ports these are also uncovered at the same time, their respective shapes being such that the in-flowing air is directed up the cylinder while the exhaust gas flows downwards on the opposite side. When the upward air stream strikes the cylinder head it is turned down again to follow the residual gases, driving them before it towards the exhaust ports. This arrangement is known as the loop scavenge system.

A more commonly employed arrangement for the diesel two-stroke, however, is the uniflow system of scavenging. In this arrangement mechanical exhaust valves are located in the cylinder head, these being opened when the effective portion of the power stroke has been completed and just before the descending piston uncovers the inlet ports. Remaining pressure in the cylinder results in the initiation of an outward flow past the exhaust valve or valves to the silencer. Immediately after the exhaust valve opens, the piston uncovers the inlet ports and air under blower pressure is admitted, the ports being so shaped as to direct the flow in an upward swirl which urges the remaining exhaust gases through the exhaust valve port. The significance of the term 'uniflow' is obvious, for the air stream does not change its direction as in the loop scavenge system, and so not only are blower power losses reduced but there is less risk of mixing and contamination with the residual exhaust gases. These advantages are obtained, however, at the cost of losing the mechanical simplicity of the loop scavenge type.

The opposed-piston type of two-stroke engine also has the advantage of uniflow scavenging. As the name implies, it is fitted with a pair of opposed pistons in each cylinder and its method of operation is fully described in Chapter 6. It also suffers the disadvantage of loss of mechanical simplicity when compared with the loop scavenge engine.

The two-stroke principle further claims attention because every piston down-stroke is a power stroke, so that the same number of impulses in unit time are provided by half the number of cylinders as compared with a similar four-stroke engine. Alternatively the same power output can be obtained from an engine of about half the cylinder capacity. Furthermore, if twice the number of impulses, each of half the intensity, can be used to produce the same power, the parts can be made smaller so that the two-stroke engine is relatively smaller, stiffer and lighter than a four-stroke of the same power. In practice, of course, the dimensions are not quite halved nor is the performance doubled, because allowance must be made for relatively greater heat and mechanical losses in the smaller unit, and for the power absorbed by the blower, not to mention the loss in volumetric efficiency resulting from the method of cylinder filling and the shortness of the inlet period.

As already explained, the explosive mixture of the petrol engine is provided by a carburettor, but in the case of the diesel engine the supply is effected by an injection or 'jerk' pump which forces a 'shot' of fuel into each cylinder in turn according to the correct firing sequence. The pump is actuated from a camshaft and the amount injected for each power stroke is extremely small; at maximum output, indeed, it may be less than 0·1 cm^3. Moreover, this charge must be metered with great and constant accuracy even while it is being variably controlled to regulate power from idling to maximum. It will be realised that efficient and economical running is largely dependent upon the fuel pump.

Multi-cylinder engines are normally equipped with cam-actuated pumps in unit monobloc form incorporating a corresponding number of elements having a common camshaft and control means but provided with individual adjustments for regulating their output and setting their timing. An equal quantity of fuel must be delivered by each element, which is calibrated separately to a predetermined rate of delivery. Timing of the start of injection in relation to engine crankshaft position must also be constant and equal on each element; the setting process is called phasing. Calibration and phasing are precision operations to be carried out only by skilled service mechanics who have the necessary specialised equipment.

A very important detail is that injection must cease cleanly and abruptly at the end of the delivery period without any trace of after-dribble from the injector; otherwise carbon deposits quickly form on the nozzle tip and excessive smoke will appear in the exhaust. The required condition is ensured by the special design of the pump delivery valves which provide for a very rapid collapse of pressure in the injector fuel pipes at the end of injection.

A familiar feature of the petrol engine is its ability to idle smoothly and run steadily at any desired speed. This is because the engine is inherently stable since the characteristics of flow through a throttle valve exercise a stabilising effect on the speed of the engine. Unfortunately, the fuel injection pump on a diesel engine does not have these desirable characteristics and it is necessary to fit a governor to control the speed.

All automotive-type diesel engines are controlled through the medium of governors in order to ensure, under any conditions of load, that speeds in excess of the predetermined crankshaft rev/min shall not be attained. It is also necessary to provide for constant and regular idling and to ensure response to sudden load application without stalling. The governor does this by adjusting fuel delivery in relation to the load on the engine, increasing it when speed falls as a result of load increase, and vice versa, thus preventing both 'runaway' at high speeds or stalling when idling. It must be understood, therefore, that the governor on the automotive diesel is not just a device to limit speed to a fixed maximum; its purpose is to provide an automatic and sensitive adjustment of fuel delivery in relation to load and speed, rather than to speed alone. On industrial engines which can be operated at constant load and speed, a simpler type of governor, responsive solely to a predetermined speed, can be used.

Modern direct-injection engines, with their high injection pressures coupled with the operating conditions peculiar to the automotive type (rapid acceleration and deceleration, in particular), impose heavy duty on the governor. Serious stresses are imposed upon the bearings of the flyweights and upon the springs. The development of the hydraulic governor was a direct consequence of the search for improved reliability.

The high cost of the hydraulic governor, however, coupled with a degree of complexity and sensitivity in servicing, directed attention back to the pneumatic type, which is simple in construction, is not easily deranged and also not a highly specialised precision job to maintain.

At one time there was a trend to extend the use of the pneumatic governor to a wider field of automotive diesels than was originally intended. It had the advantage of being directly sensitive to the torque loading on the engine, while improvements, both mechanical and pneumatic, had contributed to its capacity for stable idling control as well as satisfactory response to high-speed conditions.

At an early stage in the evolution of the high-speed diesel a variable timing device was introduced in the pump drive coupling, taking the form of a helically splined sleeve by means of which the angular relation between driving and driven shafts could be varied by manual or automatic control. This device has always been a feature of Gardner engines where it is coupled to the accelerator control. In some makes of engine the helically splined coupling is not used and the angular displacement of the shafts is effected by centrifugally actuated devices of the swinging link or inclined plane type. The fitting of variable timing devices has been discarded by some manufacturers in favour of fixed timing, but in recent years the extensive adoption of the distributor type of fuel pump, which incorporates a simple variable timing device, has now brought this refinement into much more general use.

From the above it will be realised that the fundamental difference between petrol and diesel engines is that in the petrol engine the source of heat for igniting the charge, namely, an electric spark, is generated outside the engine, and is taken, as it were, into the waiting charge at the required instant. In the diesel engine the source of heat for igniting the charge is created within the engine by compressing pure air to a degree that will initiate combustion and then injecting the fuel at the right time in relation to the movement of the crankshaft. It will have been gathered that, apart from their auxiliary features, both classes of engine are of very similar construction. But as the diesel is called upon to withstand very much greater stresses due to higher pressures in the cylinders, it has to be of more substantial construction, and is thus heavier.

By careful design and by taking full advantage of the most advanced metallurgical technique the power/weight ratio of the automotive diesel was substantially improved within the course of a very few years so that the weights of present-day engines do not appear to be capable of much further reduction.

The use of light alloys for many parts of the engine is now common practice by many manufacturers and the considerable up-rating associated with the widespread introduction of turbocharging has recently resulted in a further reduction in the power/weight ratio.

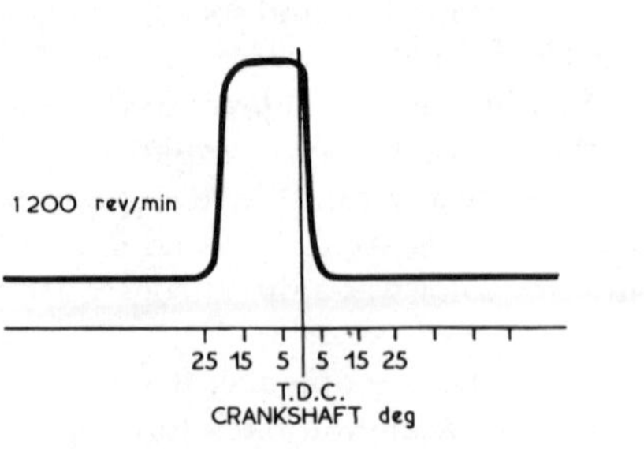

Injector needle-valve lift diagram with normal cam

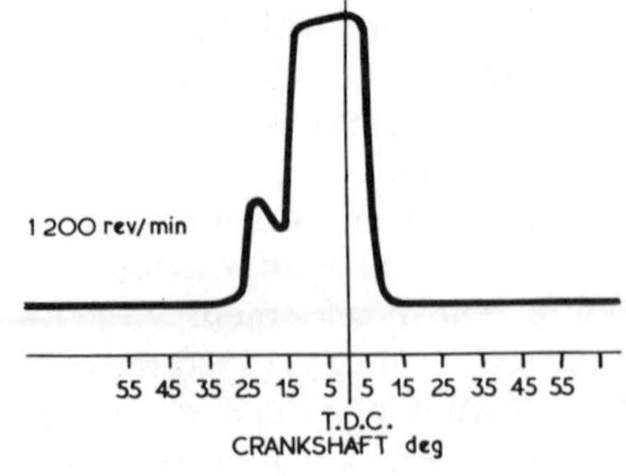

Injector needle-valve lift with two-stage cam

One of the greatest difficulties encountered in connection with the diesel engine is to eliminate what is generally known as the characteristic diesel knock, which is most pronounced when running at low speeds. This peculiarity of the diesel engine is associated with the high maximum pressures in the cylinders and with the extremely rapid rise of pressure which can occur. It is also associated with the delay period, which is dealt with in Chapter 5.

A simple analogy may be useful when considering the characteristic noise of the diesel engine in comparison with that of the petrol engine. The higher pressures and the higher rates of pressure rise in the cylinder of a diesel engine

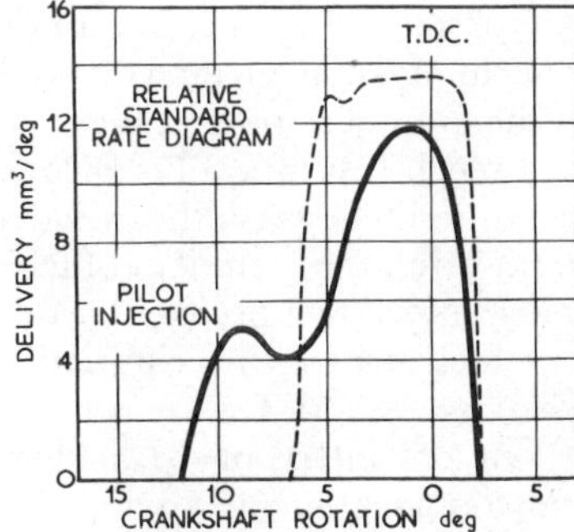

Pilot injection fuel delivery rate compared with normal injection

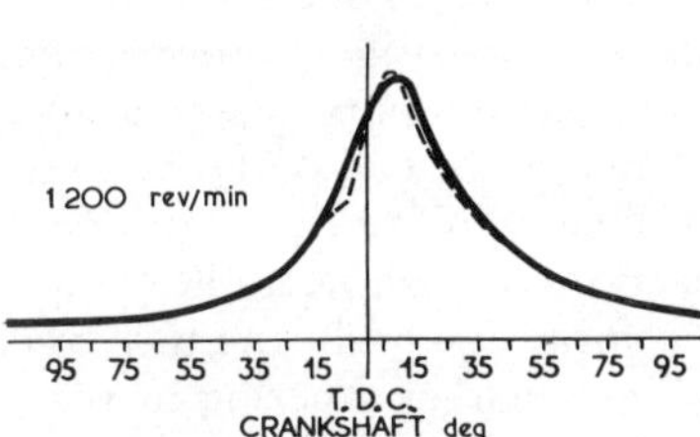

Effect of pilot injection on combustion-pressure rise in cylinder. Dotted line shows steeper and more rapid rise with normal injection

are analogous to the effect of striking the top of the piston a sharp blow with a hammer. In the case of the petrol engine the lower pressures, particularly when idling, and (provided detonation is avoided) the lower rates of pressure rise are equivalent to the effect of striking the piston a much lighter blow with a soft mallet. The heavy blow with a hard-faced hammer would excite many of the high-frequency resonances throughout the engine and cause it to 'ring like a bell', whereas a blow with a mallet would not excite so many of these resonances.

Apart from the development of pilot injection and other means of reducing the combustion noise of the diesel, much attention is now being directed to the possibility of noise reduction by changes in the design of the engine structure and its mountings, and by the application of sound-deadening materials. It is evident that the trend towards an improvement in the power/weight ratio of the automotive diesel has often resulted in an increase in the noise level, and it is encouraging to note that so much effort is now being directed towards a solution of this problem.

It will be noted that in Diesel's basic patent he stated that fuel should be introduced gradually. Much development work, both here and abroad, has been directed to this aspect of injection and what is known as pilot injection has come to be regarded as probably the final solution of the problem of diesel knock. The idea of pilot injection is to deliver a small quantity of fuel in advance of the main charge so that the latter enters a zone in which combustion is already initiated, in the belief that if the whole charge enters so rapidly as to be within the chamber before the first droplets have time to ignite, their ignition is followed by an uncontrolled combustion or, indeed, detonation of the whole charge.

Although the obvious and easier approach was to provide an increasing rate of injection, investigation carried out at the Thornton Research Centre of the Shell-Mex organisation in Cheshire indicated that the ideal arrangement was to inject a true pilot charge ahead of the main bulk with a definite break between the two phases. In the experimental set-up this was effected by the use of two completely separate injection systems on the one engine, a costly and complex method which has not yet been translated into practical application by means of a single pump. The problem is made more difficult by the need to hold the pilot injection constant at all outputs, with regulation only on the bulk charge.

As an alternative to pilot injection as a means of smoothing the combustion process, other methods have been suggested. The 'fumigation' method involves the introduction of a small quantity of fuel in the form of a spray into the air intake. In the Vigom process, suggested by the French Institute of Petroleum, a small preliminary charge is injected into the hot residual gas in the cylinder just before the end of the exhaust stroke. In both cases, the quantity of fuel is insufficient to cause pre-ignition but it is claimed that certain pre-combustion reactions occur during the compression stroke which assist with the initiation of combustion of the main charge.

Since neither tapering increase of delivery rate nor interrupted two-stage systems can be achieved without extra cost and complication, an attempt has been made in certain cases to attain a similar result by somewhat reducing the high injection pressures that have become usual, and modifying injector and delivery valve characteristics to obtain a sharp and positive cut-off at the end of the injection in spite of the weaker needle-valve spring setting resultant upon lower opening pressure adjustment. Both methods of smoothing combustion involve slight loss of fuel economy so that neither has been extensively adopted by vehicle operators who, under prevailing economic conditions, have to forgo technical improvement in the direction of smoother operation and refinement of performance where these qualities involve even the slightest increase in operating costs.

5
Fuel injection systems

In the diesel engine, injection of the fuel not only replaces the carburation system of the petrol engine but it performs the *timing* function of the electric ignition system. Thus not only are speed and power regulated by the *quantity* of fuel injected but the instant of injection in the cycle of operation has an important bearing on efficient and economical running. Satisfactory functioning, therefore, largely depends on the accuracy of the injection system.

As originally designed, diesel engines had air-blast injection, that is, the fuel was injected by air pressure of a higher intensity than the compression in the cylinder, and this system may still be found in use on a few large marine and stationary engines of pre-1939 manufacture. It had its attractions as it was capable of giving very good atomisation and penetration of the fuel, but the power required to drive the blast air compressors amounted to about 10% of the power of the engine. For many reasons it is unsuitable for small high-speed engines, and so the airless or 'solid-injection' system, in which the fuel is pumped at high pressure into the cylinders through injecting nozzles or sprayers, came into being. A separate pump and fuel-spraying nozzle is provided for each cylinder. Use of the term 'solid injection' is to be deprecated in the case of automotive diesels, however, because the distinction it implies never had significance in this field.

Injection pumps are generally grouped into an integral unit rather on the lines of a miniature monobloc engine, having hardened steel plungers which are reciprocated in the barrels or cylinders, being raised by cams and returned by springs. The pump meters the quantity of fuel and delivers it at an instant in the cycle of operations coinciding with that angular position of the engine crankshaft which is its optimum in relation to the start of combustion in the cylinder. This requires that the pump camshaft shall be timed, that is, that its angular rotation shall be in fixed relation to the rotation of the crankshaft; hence the pump drive is always positive, usually being by toothed gears, even when the engine valve timing and other auxiliary components are chain-driven. As in the case of spark ignition magnetos or distributors, the injection pump is driven at half crankshaft speed for four-stroke engines, or at equal speed on two-strokes.

In certain cases the multi-element integral pump is not used, the injection equipment being a combined pump and injector mounted on the cylinder as an individual unit. Fuel from the pump plunger passes directly through internal oilways to the injector nozzle, so dispensing with external piping.

This system is commonly used on the most popular American diesels. The combined pump and injector have not been used on British engines, but there are some British marine and industrial diesels on which a separate pump is mounted on each cylinder head connected to a normal injector by a very short pipe.

Still another form of fuel injection pump is the object of a newly revived interest, in particular for the advantages it possesses in connection with the smaller engines now being produced in increasing numbers; this class of pump is known as the distributor type and it is so called because a single pumping element is made to serve a multi-cylinder engine through the medium of a rotating distribution valve. A pump of this type is a compact unit in which the reciprocating plunger acts in phase with the firing intervals of the cylinders, the output from each successive stroke being directed to the injectors in turn. The construction is basically simple, being more compact than the multi-element in-line type, while being probably less costly to make. Easier servicing is claimed because certain important characteristics are determined by the initial settings in manufacture and are not subject to subsequent deviation or, indeed, to adjustment.

Increasing numbers of distributor pumps have come into use in the past few years, particularly on small high-speed engines, and it has been stated that they are now fitted to well over half of the engines currently manufactured in Britain and fitted with CAV fuel injection equipment.

Design and size of the combustion chamber and the position of the valves have an important influence upon efficient combustion. Other factors which all play their part are the physical and chemical characteristics of the fuel, temperature, the amount of excess air, turbulence, and compression pressure. The manner in which the fuel is injected is of equal importance, and whereas a given injection equipment may be satisfactory for one type of engine it may not suit another.

Combustion is affected by the timing, rate and duration of the injection period, the position of the spraying nozzle, the direction in which the fuel is sprayed and the injection pressure, upon which depends the degree of atomisation of the fuel and its penetration and distribution within the combustion chamber.

Thus in every case injection must be considered in relation to what occurs before, during and after the injection period, but as the conditions vary with different makes and types of engine, it will be preferable to confine the present chapter to what may be regarded as the mechanical aspect of the injection process.

Briefly stated, the function of the injection apparatus is not only to deliver extremely small and accurately metered quantities of fuel into the cylinder, but to assist in breaking the oil up into uniform particles of the smallest possible size and distributing them throughout the combustion chamber. The fuel must not be injected all at once, but over a period, commencing just before the piston reaches top dead centre on the compression stroke and ending after it has passed top dead centre; the duration of the injection period corresponds to less than one-tenth of a revolution of the crankshaft. Obviously the pressure exerted by the pump must be much greater than the initial combustion pressure in order that the fuel shall issue from the injection nozzle in a fine spray capable of penetrating the compressed air in the combustion space.

In this connection it is found that not only the size and shape, but also the length of the injection orifice in the nozzle have an important influence upon the formation of the spray. If the orifice is short, fine atomisation of the fuel is obtained, whereas with a long orifice the spray penetrates more in the required direction. A compromise between these two desirable features is usually effected. In view of the very restricted quantity of fuel required to be injected for each power stroke, the size of the orifice is extremely small, particularly in multi-hole sprayers.

For automotive engines the type of injecting nozzle employed is that in which the passage to the orifice is normally closed by a spring-loaded needle valve that is opened by the fuel itself when the required injection pressure is reached and shuts immediately it falls at the end of the injection period. Many different designs of nozzle are in use and the spray is variously directed according to the size and shape of the combustion chamber.

During the injection period the atomised fuel must be evenly distributed throughout the air in the combustion space, and so there is the use of high injection pressure, multi-hole sprayers and small injection orifices on the one hand, in conjunction with cylinder head, valve port, and piston crown designs to promote air swirl, turbulence and 'squish' on the other. These are characteristic features of direct-injection engines. In antechamber or air cell combustion chambers, atomisation and penetration are less difficult problems, so that lower injection pressure is possible, and a single-hole sprayer with a larger orifice can be used.

Even so, the condition of the fuel entering the diesel combustion chamber is very different from that in the petrol engine at the moment prior to ignition. In the petrol engine, the compressed mixture is a charge of a vaporised fuel in such intimate combination with air that it is a compressed, explosive and homogeneous vapour ready for complete and instant ignition. Injected diesel

A French-made P.M. distributor pump was fitted to a Panhard engine exhibited at the 1956 Paris show

fuel, however, in spite of its atomisation during injection, remains as droplets which are cold relative to the compressed air in the combustion chamber. The fuel and air are in process of mingling so that the mixture is not yet homogeneous, and furthermore it is the heat of the air which must ignite the fuel.

Now, although ignition may occur quite soon after the start of injection, it is limited to the surface of the droplets, which 'boil off' into vapour and become ignited by the temperature of the surrounding air. During this phase of ignition the temperature is raised still further until a cumulative combustion of the entire charge of fuel results. Obviously the rate and duration of combustion can be influenced by the size of the droplets, as the smaller droplets have a higher surface-to-volume ratio, thus presenting a greater area to the surrounding air. Also there is a time lag, known as the delay period, between the start of injection and the first ignition of the fuel, when certain pre-combustion reactions occur. This is influenced by the temperature of the air in the cylinder when injection starts, but it is also considerably influenced by the physical and chemical properties of the fuel. The delay period can cause various disturbing features, notably diesel knock, rough idling and difficult starting.

Numerous theories have been put forward to explain the delay period, but it is generally accepted that the physical and chemical properties of the fuel are of paramount importance. Thus, the viscosity and surface tension can influence the size of the droplets and the volatility can influence the rate at which the surface of the droplets can boil off into vapour, but it is the chemical composition of the fuel which determines the spontaneous ignition temperature and the ignition lag. So the quality of the fuel as well as the characteristics of the fuel injection equipment can both have a considerable effect on the ease of starting and the smoothness of running of a diesel engine. One method which

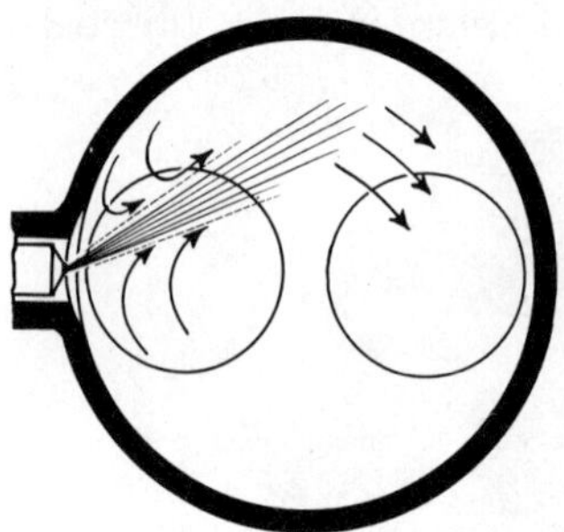

An instance of fuel injection tangential to an air swirl induced by inlet-valve design. In other examples the fuel may be sprayed through more than one orifice in the nozzle

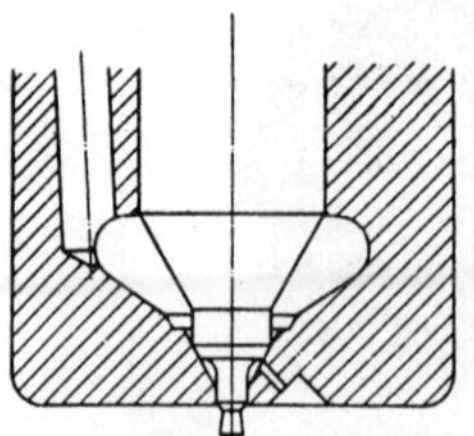

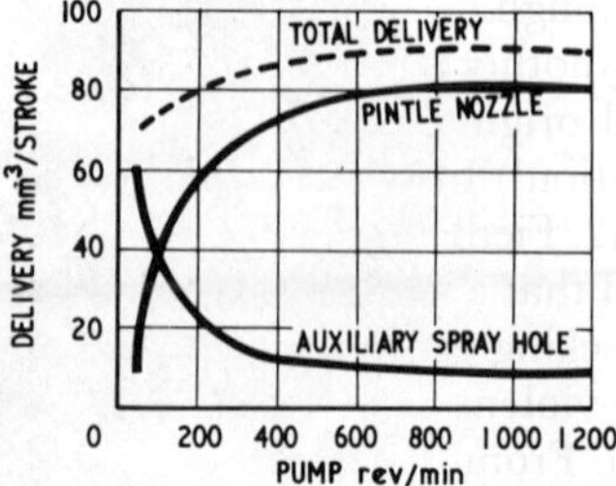

CAV Pintaux nozzle showing auxiliary spraying hole beneath valve seat; the curves indicate its performance characteristics

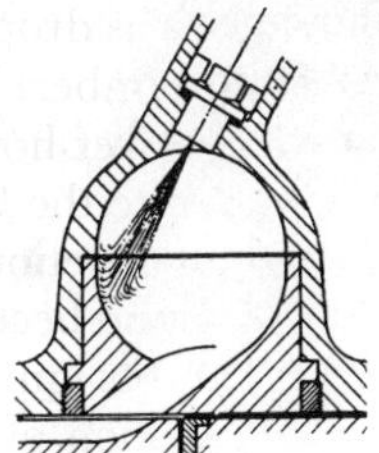
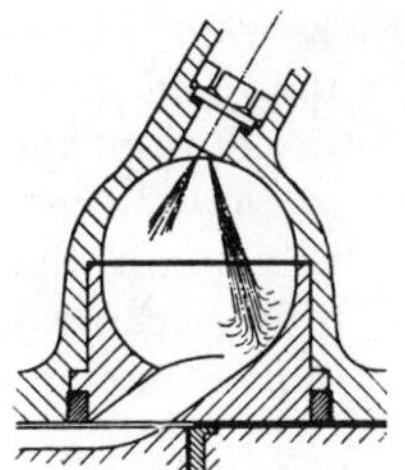
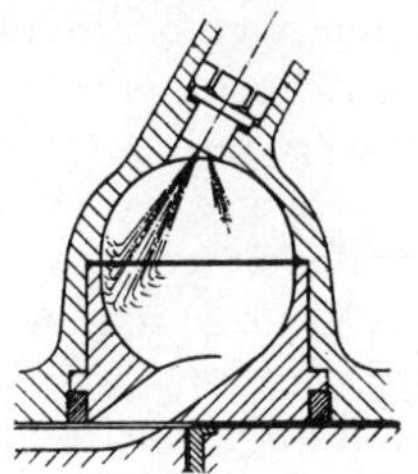

(Left) *Normal pintle-nozzle spray in Ricardo Comet chamber.* (Centre) *CAV–Ricardo Pintaux nozzle operating at low speed with bulk of fuel injected towards centre of cell.* (Right) *The same at high speed with small spray towards centre and bulk injection following normal pintle-nozzle characteristics*

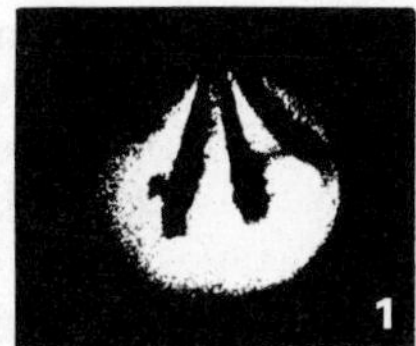

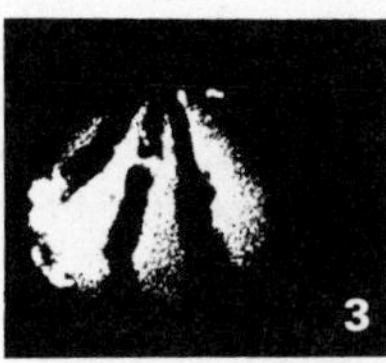

Photographs from Fiedler's cine film of a four-hole spray in the combustion chamber. White specks in (3) suggest early combustion of dissociated hydrogen

has been successful in limiting diesel knock is that of injecting a small quantity of fuel as a pilot injection as described in Chapter 4.

Considerable attention was devoted to pilot injection, notably by the Atlas concern of Sweden, in the mid-1940s; their adaptation of the idea was directed towards the possibilities of two-stage injection whereby the first part of the injection is at a relatively lower pressure, followed by the main bulk at higher pressure; this is effected by a double-lift cam on the pump camshaft. The object of two-stage injection is to initiate combustion in an orderly manner without the violent detonation or diesel knock, and then to sustain the subsequent combustion more nearly according to the *constant pressure* cycle of the original Diesel concept. The further development of the Atlas system in this country was in the hands of CAV Ltd. Applying the same principle in a rather different way, a special CAV–Ricardo injector was developed for use with the Ricardo Comet combustion chamber. Known as the Pintaux nozzle this is a pintle-type sprayer with an additional side orifice so arranged that a portion of the injected fuel is directed to the centre of the air cell where the highest temperature exists, while the bulk of the fuel is injected tangentially to the air swirl in the normal way. This injector facilitates cold starting and smooths combustion although no claim for pilot injection is made.

Another contribution to the theory of compression ignition of great interest and originality was embodied in a paper ('A New Theory of Diesel Combustion') by Max G. Fiedler before the Franklin Institute of Philadelphia in 1942. Fiedler stressed the point that diesel knock was analogous to pre-ignition and that a study of indicator diagrams taken on diesel engines running in the knocking condition showed characteristics very similar to those obtained from the violent detonations of an internal combustion engine run on pure hydrogen. From this he deduced that the hydrogen of the hydrocarbon fuel oil was dissociating from the bulk of the fuel injected and igniting in advance of the remainder, so causing the diesel knock. This theory was supported by the evidence of cine films taken in the working cylinder and showing that the spray remained cohesive much longer than had previously been supposed and that there were indications, in the form of bright specks of flame, of early burning in a heterogeneous way at points remote from the nozzle.

Fiedler's theory did not attract the notice that his highly stimulating paper deserved; but it is worthy of study because it may well present a line of approach along which further improvements in the control of diesel combustion might be obtained.

To appreciate the difficulties of designing and constructing satisfactory fuel injecting equipment for high-speed diesel engines it is necessary to consider the exacting mechanical requirements that have to be fulfilled.

Having regard to the fact that oil fuel is to some small extent compressible and that the delivery pipes connecting the pump to the injectors are similarly expansible, the metering of minute charges of fuel at high delivery frequency demands great accuracy in the components and extreme rigidity in mounting. Theoretically the delivery pipes should be short and of precisely equal lengths. It is also important that the sizes of the pumps and passages in the delivery system are no larger than necessary and that the surfaces over which the fuel passes are smooth and free from cavities where air can be imprisoned and so increase the elasticity of the fuel in the system. Fine filtering arrangements are essential in the fuel feed to the injection pump in order to remove foreign matter liable to cause damage to the highly finished surfaces of the injection equipment.

It is very important that the action of the pump should be such that the injection pressure at the nozzle is reached very quickly; that is, while the crankshaft turns through a fraction of a degree, and that at the end of the injection period it be reduced even more rapidly to zero. The latter requirement is particularly important to avoid any after-dribbling of fuel from the nozzle with consequent fuel waste, smoky exhaust, carbon deposit and choking-up of the spraying orifices. It is necessary, also, that the pressure be maintained evenly during the injection period. These conditions, moreover, must be fulfilled at all speeds of the engine and be the same for each cylinder.

Considerable variety was seen in the many designs of fuel injection pumps put forward in the development stages of the modern diesel, but two main principles were generally followed. Regulation of the quantity of oil delivered was effected either by variation of the length of pump plunger stroke, or by the use of controllable bypass valves or spill ports in conjunction with a constant stroke. The latter type is now widely used and its distinguishing feature is that on each suction stroke the plunger draws in a greater quantity of fuel than is required to be delivered for maximum load, the actual amount injected being controlled according to requirements by bypass or spill valve regulation; that is to say, during the delivery stroke of the plunger, communication is established between the delivery and suction sides of the pump so that at a desired point during the injection period the delivery pressure is suddenly reduced and injection ceases. This is usually effected by the plunger itself acting as a slide valve which uncovers relief ports in the cylinder barrel to terminate delivery in the manner described; the plunger or the cylinder barrel is rotatable to provide the necessary control.

A radical departure from this principle is featured in the distributor pumps which have become popular in recent years. In this case the quantity of fuel is controlled by a throttle valve on the suction side of the pump which limits the travel of the plungers during the suction stroke.

Many different designs of spraying nozzles have also been introduced, their construction and action being varied to suit different types of engine and forms of combustion chamber. They are of either the open or the closed type, the latter being provided with some form of valve for closing the injection orifice until it is opened by the fuel pressure for the purpose of injection. The closed type is invariably used in automotive diesels. Nozzles are of either the multi-orifice or single-orifice type, the former being employed to give better dispersion of the fuel in direct-injection engines.

The closed type of nozzle requires some means of eliminating any accumulation of air in the delivery system. This may take the form of a vent cock that can be opened by hand, although this fitting was generally discarded as improved injectors and non-frothing fuel came into use. Venting is rarely necessary today as an ordinary running attention but it must be done if a connecting union in the delivery pipe becomes loose, or after reassembly following repair or replacement. In such circumstances, in the absence of a venting cock or plug, the union on the injector is not finally tightened until the pump has been operated a sufficient number of strokes to deliver unaerated fuel after driving out all air from the pipe, when the union can be firmly tightened.

Although in the early development of the automotive diesel many fuel pump designs were put before engine manufacturers, the Bosch equipment was the first and only type to be available in quantity. It will be recognised that the purely technical difficulties of producing such a high-precision component on a commercial scale were considerable, and in overcoming these the Robert Bosch AG of Stuttgart undoubtedly led the world; as has been mentioned earlier, that concern did indeed make possible the rapid progress of the modern automotive diesel engine.

In 1933 the manufacture of pumps and injectors to the Bosch design was taken over in this country by CAV Ltd., of Acton, London, and considerable further development has since taken place. Other British makes of pump and injector equipment supplied extensively are Simms and Bryce. Similarly in the USA equivalent diesel-injection equipment is produced by the American Bosch Co. But there are also other types of fuel pumps in extensive use in the USA which so far have not been introduced here—indeed, they are distinctive types which appear to be peculiar to some of the most extensively used American diesels.

The various British pumps are similar in principle to the Bosch constant-stroke arrangement with variable spill control for power regulation; detail variations naturally are incorporated according to the particular designer's ideas on the factors necessary to increase efficiency and reliability. However, since the CAV pump is in such general use it will be described first.

The CAV pump is a self-contained unit embodying an operating camshaft which is driven from the timing gear of the engine. A governor is fitted as an integral assembly, either on an extension of the camshaft in the case of the mechanical-centrifugal or hydraulic types or on the upper part of the end casing when the pneumatic type is used. The mechanical governor is the type hitherto most commonly fitted to large engines, while the pneumatic governor has been applied mainly on small- or medium-capacity engines running at speeds above the general average.

Apart from these subsidiary variations the pump itself is a standardised design that has not been subject to any great change since its introduction and the general description applies to all CAV in-line equipments in current use irrespective of age. The N-type pump introduced towards the end of 1948 was somewhat reduced in size and many mechanical improvements were incorporated, particularly an internal filter unit, larger bearings and improved adjutments for phasing and calibrating; also a diaphragm feed pump was fitted. The NN-type, a later version, includes still further improvements and in Stage II form, it is suitable for dual fuels (petrol and/or Derv), while the Stage III form is suitable for application to more highly rated engines. Both these modified forms include improved sealing and lubrication features.

For each cylinder of the engine a separate injection system is required, and therefore the pump is a multi-cylinder unit incorporating the requisite number of individual pump elements (up to eight) in a single casing, each element comprising a steel barrel and a steel plunger, ground to a very high degree of accuracy. Operated by a camshaft mounted in ball bearings, each plunger is

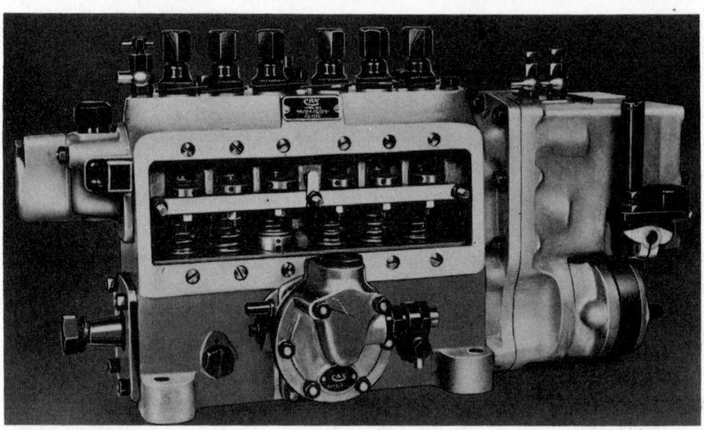

CAV N-type injection pump fitted with hydraulic governor and fuel lift pump

provided with a roller tappet carried in a suitable guide and held down by a coil spring on to its respective cam. Near the upper end of the barrel are drilled two ports opening into a suction chamber which is common to all the elements. Above and communicating with the pump barrel is a spring-loaded delivery valve, from which connection is made by steel tubing to the injection nozzle. The fuel is admitted into the suction chamber through the connection which is to be seen on the left in the accompanying illustration of the partly sectioned four-cylinder pump unit.

The manner in which the quantity of fuel delivered per stroke of the plunger is varied according to load requirements can be gathered from the sectional illustrations (p. 38) showing the plunger in different positions of operation. From these it will be seen that on one side of the plunger there is a vertical channel leading from the top edge to an annular groove, the upper edge of which forms a helix. The plunger can be partly rotated by means of a control sleeve which has slots engaging with lugs at the lower ends of the plunger and fitted with a quadrant which meshes with a control rack.

In the first figure of the diagram of the injection-pump cycle mentioned the plunger is shown at the bottom of its suction stroke, and commencing the delivery stroke; the two ports in the barrel are uncovered so that the barrel is filled with fuel. As the plunger rises on its delivery stroke the fuel in the barrel is displaced and is forced back through the ports until these are entirely closed by the plunger. What fuel remains is forced upwards through the delivery valve to the injecting nozzle. So long as the ports are kept closed by the plunger, injection of the fuel is continued, but, as shown in the second sectional illustration, before the plunger reaches the top of its stroke the helical edge of the annular groove has partly uncovered the port on the right. The fuel above the plunger is then free to flow down through the vertical channel and annular groove and through the port into the suction chamber. The pressure is thus released so that the delivery valve can return to its seat and injection ceases.

In the third diagram the plunger is again at the end of its suction stroke, but has been turned slightly by means of the rack and quadrant referred to, so that on the next delivery stroke the helical edge of the annular groove commences to uncover the relief port earlier, as shown in the fourth diagram, with the result that a smaller quantity of fuel is injected. When the plunger is turned to the position shown in the fifth diagram, throughout the whole delivery stroke the fuel is free to be passed down the vertical channel and through the relief port into the suction chamber so that no fuel is injected, and the engine stops. Thus it will be seen that by movement of the control rod, which is connected to the governor and to the accelerator pedal, the amount of fuel injected can be varied to suit load requirements. The maximum output of the pump is obtained when the plunger is in the position shown in the first diagram.

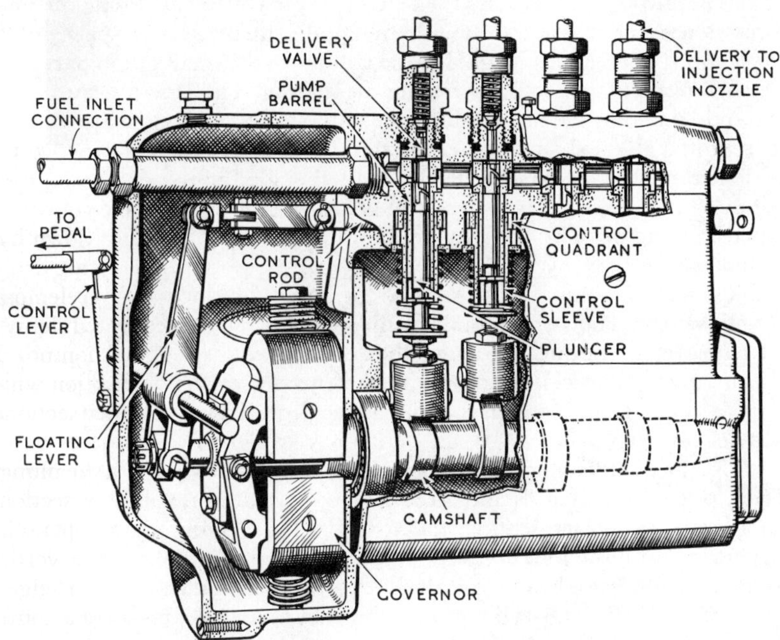

CAV fuel pump and centrifugal governor in part-section

Reference has already been made to the importance of a quick cut-off at the end of the injection period so that there is no after-dribbling of fuel at the nozzle. This is guarded against by a special form of delivery valve, the function of which is to release the pressure in the fuel pipe between the delivery valve and injection nozzle immediately the pump pressure drops.

The delivery valve is mitre-faced, but is provided with a long cylindrical extension, which fits in a cylindrical guide and has an annulus dividing it into

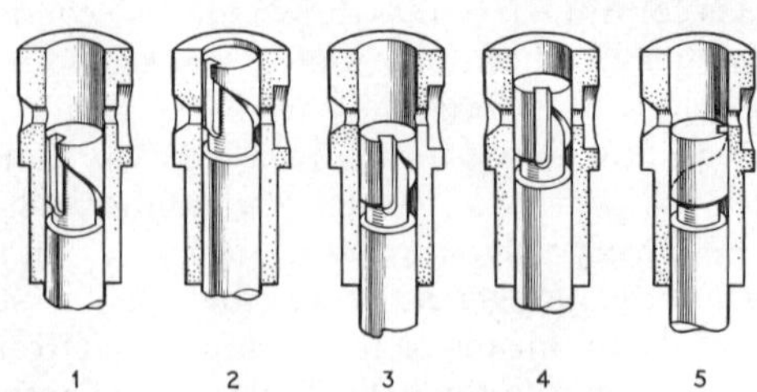

Injection cycle of CAV fuel pump: (1) *Commencement of delivery stroke* (2) *Termination of fuel injection under full load* (3, 4) *Commencement of delivery stroke and termination of injection under partial load* (5) *Position of plunger for stopping engine*

two parts, the lower portion being of cruciform section. The upper part forms a small piston very accurately ground to fit the cylindrical guide below the valve seat, which is also accurately ground. On the delivery stroke of the pump the pressure lifts the delivery valve until the oil can escape through the longitudinal grooves and over the valve face on its way to the injection nozzle. Immediately the pressure is released in the pump barrel the delivery valve is returned to its seat both by its spring and because of the great difference of pressures in the pump barrel and the delivery pipe. During its closing the small piston, in passing down the cylindrical guide, increases the space in the delivery pipe (by an amount equal to the volume of the small piston part of the valve) before the valve actually reaches its seating. Thus the pressure in the pipe is suddenly reduced so that the valve in the injector nozzle snaps down on to its seat and the fuel spray is terminated without dribble. In practice it is usual to arrange for the volume displaced by the small piston part of the valve to be such that a residual pressure, well below the injector closing pressure, is retained in the pipe. This has been found to improve the precision of the fuel injection process.

A phenomenon only recognised in the later developments of the fuel injection system is known as secondary injection and it is caused by the rapidity of pressure drop in the fuel pipe which the delivery valve is designed to produce. The consequent snap closure of the injector needle valve which is produced initiates a reflex wave through the fuel in the pipe. This is returned from the closed delivery valve and can be sufficiently powerful to bring about a momentary re-opening of the injector. A special form of delivery valve is used on some fuel pumps to eliminate this undesired action. The valve is a light disc closing on to a spring-supported seating, the assembly thus having limited flexibility sufficient to damp the reflex wave.

Mention should also be made of the CAV-type BPE flange-mounted pumps supplied in 1-, 2-, 3-, or 4-cylinder units for mounting on cam-boxes provided by the engine makers. An example of this type of construction is found on Gardner engines.

The idling and maximum-speed governor already mentioned as an essential component of injection equipment, is enclosed in a housing integral with the pump and is shown at the left in the drawing of the complete pump on page 37. It comprises two spring-loaded and centrifugally actuated weights mounted on an extension of the camshaft and operating a pair of bell-crank levers connected to the lower arm of a floating lever, the upper end of which is coupled to the pump control rod. The floating lever is eccentrically mounted on a shaft fitted with a lever connected to the accelerator pedal. The action of the governor weights as the speed increases is to pull the control rod in the direction which reduces the fuel delivery. The operation of the accelerator pedal is independent of the governor, for when depressed it turns the eccentric shaft, moving the floating lever, and therefore the control rod, so that more fuel is delivered irrespective of the position of the governor weights. Each of the latter is controlled by an outer spring for regulating the idling speed and a stronger inner spring for controlling the maximum speed.

Under idling conditions the weights bear only against the outer springs, but when the engine speed, and therefore the centrifugal force, is increased, further outward movement throughout the normal speed range is prevented by the stronger inner springs, so that the engine speed is controlled only by the accelerator pedal. When the latter is further depressed, however, so that the pump delivers more fuel than the engine needs for the load, the speed, tending to exceed the predetermined maximum, increases the centrifugal force sufficiently to compress the stronger springs, and in so doing the weights act upon the floating lever and control rod so that the engine speed is held down to the predetermined maximum.

A centrifugal governor differing greatly in principle from the maximum and minimum CAV type is fitted to Gardner engines on which CAV pump elements are used on a camshaft made by the engine builders; it is operative throughout the whole speed range, the accelerator pedal being essentially a

CAV anti-dribble delivery valve. This device releases the pressure in the fuel pipe between delivery valve and nozzle immediately the pump pressure drops. Because of the sudden reduction in pressure the valve in the injector nozzle snaps down on its seat, and the fuel spray is terminated without dribble

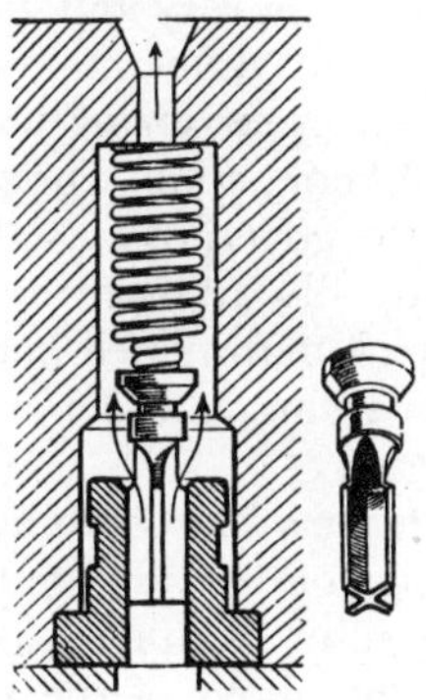

speed control acting through a device which regulates fuel delivery according to torque loading. In action the movement of the bob weights against a spring coaxial with the governor shaft is transmitted to the pump slider bar which closes down the fuel delivery. The accelerator control does not at any time directly move the slider bar; it loads the above-mentioned spring, so delaying the movement of the governor weights until a higher engine speed is reached; thus at low speeds when the accelerator is not depressed the maximum delivery is made but, if the load is light, the speed rises quickly and the governor weights fly out at once against the practically unladen spring and reduce the fuel. Depression of the accelerator increases the pressure on the spring, which delays the governor action until the speed rises sufficiently to enable the weights to overcome the resistance. At full accelerator pedal depression the spring is compressed to its greatest resistance and the engine, at full fuel delivery, must overcome its load and its speed must rise to maximum before the governor comes into action to hold down the maximum permitted rev/min.

It will be seen that at all times the tendency is for the engine to receive maximum delivery, which is then reduced as the speed rises. The function of the accelerator is not to increase delivery but to delay the action of the governor

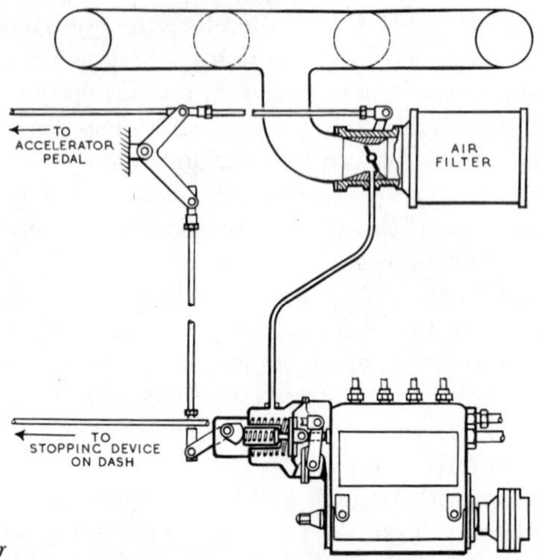

CAV pneumatic governor

in relation to rising engine speed. Thus full depression of the pedal does not increase fuel delivery but permits it to remain without restriction at any engine speed up to the maximum rev/min. Intermediate accelerator position loads the governor spring less heavily, so that rising engine speed results in earlier governor action. The accelerator pedal, therefore, determines the speed at which the governor acts although until that limit is attained fuel delivery is automatically adjusted to the torque requirement of the moment. If the accelerator is maintained in a set position, delivery is strictly proportioned to meet the varying load demand on the engine while the corresponding speed is held steady.

Another form of governing is the pneumatic system which derives from a Bosch design of the mid-1930s. In various British versions of this arrangement the mechanism comprises a throttle valve unit which is flanged on to the induction manifold of the engine and a diaphragm-control unit which is mounted on the injection pump end plate. The former contains a venturi at the narrowest part of which is a butterfly control valve (controlled by the accelerator pedal) and a pipe connection to the diaphragm housing. Adjustable stops locate the control valve in the idling and maximum-speed positions. The diaphragm is connected to the control rod of the injection pump, and a coil spring tends to hold the rod in the full-load position; this action is opposed by the vacuum which is applied to the diaphragm from the throttle valve unit. Acting also on the control rod is a lever by which the rod can be moved manually to stop the engine by shutting off the fuel delivery.

When the driver takes his foot off the accelerator, the control valve closes so that the vacuum prevailing in the venturi overcomes the pressure of the diaphragm spring and draws the control rod away from the full-load position until the engine runs at the desired idling speed. If the control valve is slightly opened by pressure on the accelerator pedal, the vacuum in the venturi diminishes so that the coil spring forces the control rod towards the full-load position, and the engine speed rises until the depression in the venturi so increases that the rod is again drawn in the opposite direction. Thus engine speed is regulated according to the position of the control valve. If the latter is fully opened the maximum speed of the engine is maintained, and if this is exceeded a higher vacuum is produced in the venturi, so that the control rod is drawn towards the reduced fuel position until the engine speed again falls to the predetermined maximum.

Another pneumatic governor coming into wide use is that developed in connection with the Simms SPE-type pump. Like the CAV, it depends upon a diaphragm in a closed chamber subject to manifold depression. An important feature of this device is that two pipes are led from the vicinity of the controlling throttle, one from its atmospheric side and the other from the engine side, the object being to maintain balanced conditions and to eliminate surging while idling.

Hydraulic governing has been constantly before the minds of diesel engineers; indeed, one of the earliest successful road transport oil engines, the Saurer, had a hydraulic governor. From time to time other governors of this type have made an appearance without establishing themselves permanently.

Although hydraulic governors can give excellent speed control, they are not used as extensively on automotive engines as might have been expected. They are somewhat complicated instruments calling for high-precision

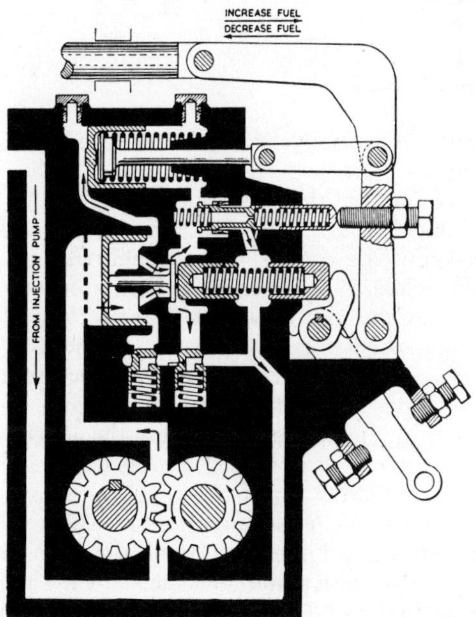

Oil-circuit diagram of CAV hydraulic governor

methods of manufacture, which tends to a high initial cost. Moreover, for their dismantling or adjustment, they require considerable skill and conditions of great cleanliness.

The CAV hydraulic governor has been in service on high-speed engines for many years and its operating principle can be taken as typical. A gear pump driven by the injection pump camshaft draws fuel from the pump gallery and delivers it under pressure to a spring-loaded piston fitted with a metering orifice. The pressure acting on this piston is therefore a function of engine speed and the movement of the piston against its spring is used to open an amplifier valve which admits the oil to a spring-loaded servo piston coupled to the control rack of the fuel injection pump. Speed control from the accelerator pedal is effected by a lever which applies additional load to the orifice piston spring. Various additional refinements are fitted such as an idling valve to give greater governor sensitivity at idling speed and slight modifications can enable it to operate as an all-speed governor, a two-speed governor or as a single-speed governor. A special version for close speed control on generating sets is also available.

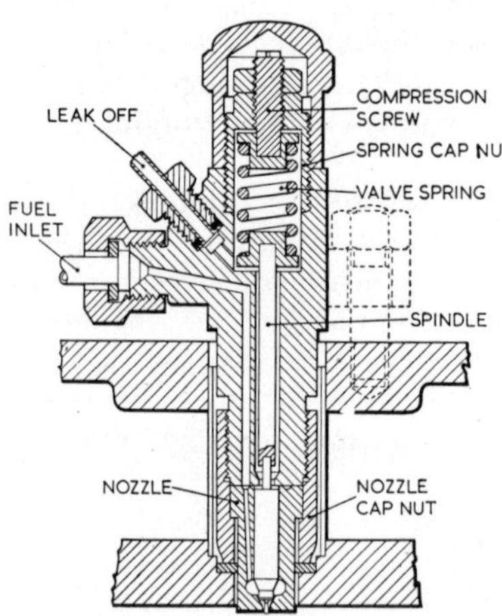

Section of CAV injector. The nozzle enters the combustion space through a copper sleeve in the cylinder head water jacket

It must be pointed out that the governing requirements vary considerably as between road vehicles and marine and industrial units. For the former a steady idling speed must be established and held whenever the accelerator control is released, while a predetermined maximum speed must not be exceeded irrespective of the control position and the torque loading. Between idling and maximum speeds the output of the engine may be directly controlled by variation of fuel delivery by the driver through the accelerator pedal linkage. Marine and industrial engines, on the other hand, must remain steady at any speed selected by the operator, therefore the governor must be of the 'all speed' type, sensitive at any setting within its range. Pneumatic and hydraulic governors are readily adjusted to these conditions and so are many forms of centrifugal governor, especially those in which the manual setting acts on the governor spring without direct connection to the pump control rod.

The CAV injector or nozzle is of the closed needle-valve type, and its general construction is shown in the sectional drawing on this page. It will be seen that the nozzle valve is held on a conical seating by a spindle and coil spring, the compression of the latter being adjustable by means of a screw and lock nut. The fuel inlet connection is shown on the left, and the fuel is led through a drilled passage leading down to the bottom of the nozzle holder, and communicating with an annular semicircular groove in the upper face of the nozzle. Thence a drilling in the nozzle body leads the fuel to a small chamber adjacent to the tapered portion of the needle above the valve seat. The nozzle is secured in place by a cap nut, and its upper face, as well as the face of the holder against which it is tightened by the cap nut, is finely ground to ensure a fuel-tight joint.

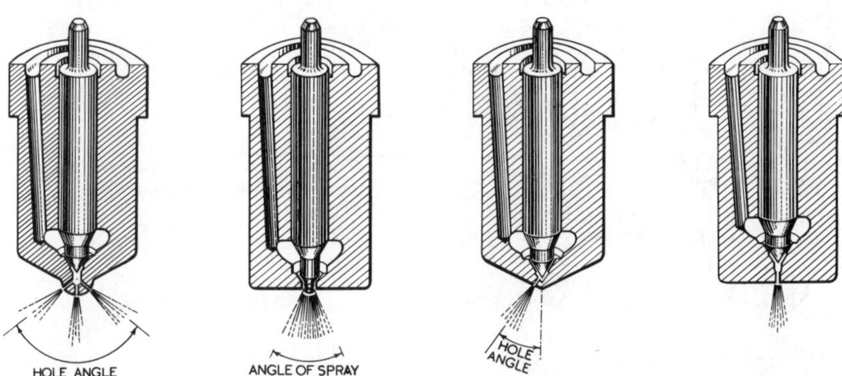

Types of injector nozzles: multi-hole; pintle; single-side-hole; single-central-hole

The spring compression is so adjusted that the valve is lifted from its seat when the required injection pressure is reached, and closed when it is reduced at the end of the injection period. Any slight leakage of fuel which may accumulate above the valve passes into a chamber surrounding the spring, and, by means of a pipe connection, it is led back to the pump suction chamber or to a tank.

The manner in which the fuel is sprayed depends upon the design of the combustion chamber. Thus injectors are supplied with different types of nozzle, examples of which are illustrated.

The pintle type of nozzle, suitable for pre-combustion and air cell engines, has an extension of the valve forming a pin or pintle which protrudes through the mouth of the nozzle body. By varying the size and shape of the pintle, a cone of spray from 4° upwards can be provided according to requirements.

Certain engines, usually of the pre-combustion chamber type, require nozzles with modified spray characteristics in order that they can produce a stable performance when idling. This is obtained by the use of what is known as a delay nozzle in which, by a modification of the pintle, the rate of injection is increased towards the end of the delivery, the effect of this being briefly to lengthen the periods of injection at idling speeds without affecting combustion at higher speeds.

The single-hole nozzle has a central orifice closed by the valve; the diameter of this can be of any required size from 0·2 mm (0·007 in) upwards. A variation of the single-hole type is known as the conical-end or side-hole type. One hole only is used, but it is drilled at an angle to the vertical centre line of the valve as required. This type is now rarely encountered and may be regarded as obsolete in automotive practice.

Multi-hole nozzles can have any number of holes drilled through a central projection of the nozzle tip, but the number is usually small. The holes are mostly arranged radially in a single line with even pitch about the axis of the nozzle, the number, size and angle varying according to the needs of the engine.

In 1934, after some five years of intensive development, the first Simms fuel injection equipment appeared on the market. The earliest model was the

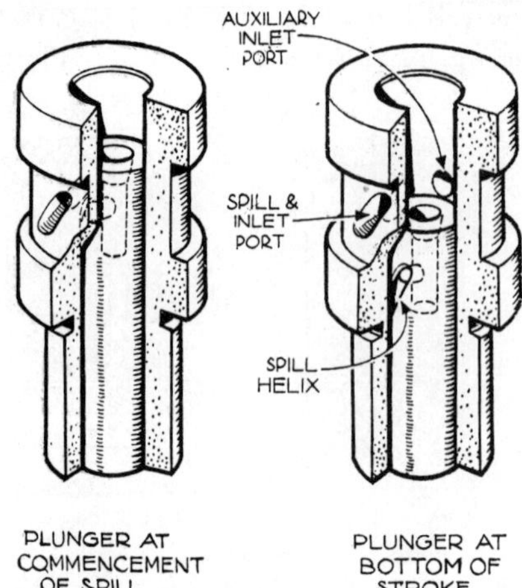

Showing the operating principle of the Simms axial-hole pump plunger

Uniflow pump, so called because the oil flow through the pump was constantly in an upward direction. Fuel injectors were also marketed by Simms and were used on Leyland engines for a time.

After further pioneering work, Simms introduced the PA model in 1938, later known as the SPE-B series, and this model was used on the engines of tanks and military vehicles during World War II. The pump had a number of special features, many of which have been retained in successive models until the present. For example, the upper half of the body of the pump, containing the working elements, can be removed without disturbing the camshaft. Also, the plunger design is unusual in that a central hole in the top of the plunger communicates with a helical groove cut in the side of the plunger a short distance from the top. At the end of the delivery period, this groove registers with the spill and inlet ports drilled in the barrel, the quantity of fuel delivered being controlled by the usual method of plunger rotation. This design has the advantage of an unbroken surface around the tip of the plunger, where leakage and wear are most severe. In the PA model, the rotation of the plungers for quantity control was by a rack-and-pinion mechanism, the pinions being split and clamped by hardened steel screws to permit individual adjustment.

Further development of the PA pump led to the Simms SPE-B series which continued in production for many years, and in this model the rack-and-pinion control mechanism was abandoned in favour of a dog-and-lever device now used on all Simms pumps.

Further development work led to the production of a small Simms pump suitable for engines of less than 1·25 l. per cylinder known as the SPE-A series, and in 1950 these pumps were made a standard fitting on the engines of Fordson tractors as well as a number of commercial vehicle engines.

The SPE-M series, also known as the Minipump, was introduced in 1959. It retained many of the features of the earlier designs, but it was more compact, the most notable change being in the method of production of the pump body containing all the pumping elements and fuel passages, which was machined from steel bar instead of the aluminium alloy castings used on earlier models. This made possible the closer spacing of the elements giving a reduction in the overall length of the pump and allowing the use of a sturdier camshaft. Mechanical governors are now built into the cambox casing instead of the earlier practice of attaching a separate governor casing to the end of the pump block.

The Simms SPE series pumps are now produced in two sizes known as the Minimec and the Majormec. The SPGE-M Series Minimec pumps are manufactured in 3-, 4-, and 6-cylinder units suitable for engines up to 1·5 l. per cylinder, while the Majormec series is available in 4-, 6-, and 8-cylinder versions suitable for engines of between 1½ and 3½ l. per cylinder. Simms in-line pumps are currently fitted as standard by a large number of engine manufacturers, and it is estimated that there are over two million mini-type pumps in use.

Another well-known make of fuel injection pump is the Bryce, made by Bryce Berger Ltd., of Gloucester. This is of the usual constant-stroke, spill-control type, but it embodies a number of exclusive features to facilitate adjustment and to reduce wear. The name has been associated mainly with single-element, flange-mounted pumps for large industrial engines, but they also make a range of small single-element pumps with plunger diameters between 5 and 9 mm (0·2 and 0·35 in) fitted as standard on a number of small industrial engines, including those manufactured by Petters.

Simms PA pump showing the detachable element block which is also a feature of the latest SPE model

A range of Simms atomisers is also available. Special attention has been given to cooling; the needle guide is well up inside the injector body so that it is surrounded by the cylinder head water jacket, while the lower end of the needle valve and the nozzle are cooled by the fuel which passes down a clearance around the needle. The atomiser is thus working under more favourable conditions than where the needle guide is located in the nozzle itself.

Apart from the injection nozzles made by the various fuel-pump manufacturers, certain engine makers produce their own injectors, notably Gardner

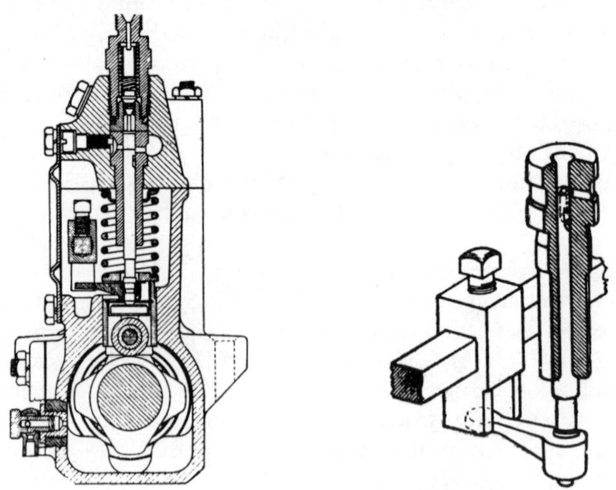

(Below) *Simms SPGE-M Series (Minimec) fuel injection pump with centrifugal governor and fuel feed pump. The section* (above left) *shows the detachable element block while the accompanying sketch* (above right) *shows the dog-and-lever method of control*

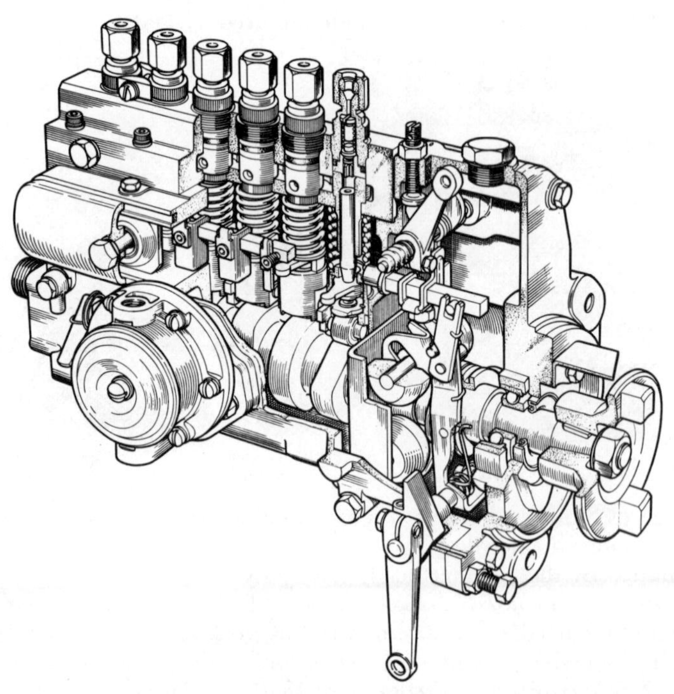

and Leyland. The direct-injection Gardner engine was one of the first units fitted with the multi-hole sprayer, and its own-make injectors were fitted to the original high-speed type introduced in 1929. Gardner injectors are of the four-hole non-adjustable type with remote valve seat, that is to say, the needle-valve setting is up inside the nozzle and well away from the tip, which projects into the combustion space and is subject to maximum heat. The top of the body is closed by a breech plug having a guide rod which enters the hollow valve seating is up inside the nozzle and well away from the tip, which projects hollow valve.

The fuel enters an annular space around the lower portion of the valve, and when the pressure exceeds that of the spring (usual injection pressure of about 155 kgf/cm^2 or 2 200 lbf/in^2) the valve is lifted and injection commences. The construction is very simple, and when taken apart for cleaning cannot be reassembled incorrectly.

Venting of the delivery pipe and the testing of injectors is facilitated on Gardner engines by the provision of hand priming levers on each pump

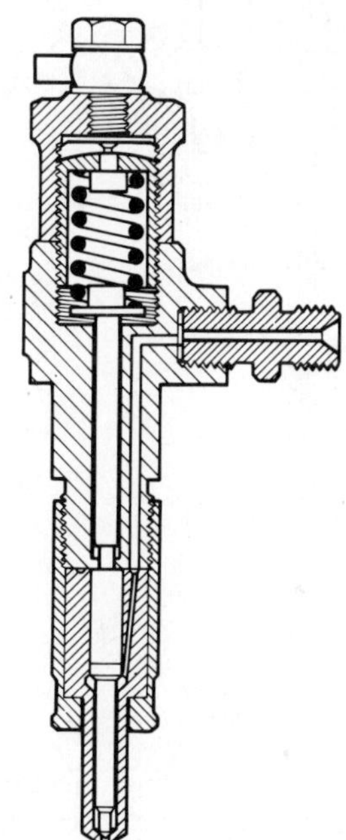

A Simms fuel injector

element, a feature incorporated in the Gardner camshaft and base on which the CAV pump elements are mounted.

The Leyland concern used CAV injectors in their earliest engines, but subsequently introduced an alternative of their own make. This was of very simple construction, of the single-hole type with remote valve seat. An interesting feature of this injector was that although of the single-hole type it was made for a direct-injection engine, thereby indicating that the use of multi-hole sprayers with direct-injection combustion chambers was not invariable in the early period of automotive diesel history. No external adjustments were provided on the injector, the working injection pressure of some 160/170 atmospheres being determined by the needle valve spring which was pre-set between certain small limits by the use of adjusting shims at its abutment against the leak-off plug at the top of the body.

With their subsequent adoption of the toroidal-cavity type of direct-injection combustion chamber, Leyland developed a new four-hole injector. It is of similar construction to the previous single-hole type, being non-adjustable apart from the shim-setting adjustment of the spring during assembly. An excellent feature is the simple and easily cleaned edgewise filter incorporated in the fuel-pipe union.

Much of the same process of evolution was apparent in the USA as in Great Britain and in Europe. In the early period the pioneer diesel-engine makers

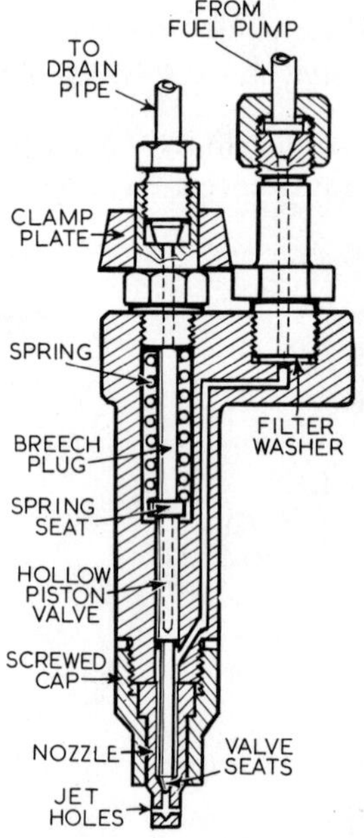

Gardner injector. This is of the four-hole, non-adjustable type with remote valve seat. It is essentially simple in construction and is very easy to reassemble after cleaning

Leyland injector with multi-hole nozzle. The fuel pipe union incorporates a simple and easily cleaned filter

produced their own pumps and injectors until the American Bosch Corporation came into being and made the necessary diesel apparatus available in standardised production form applicable to all types of engine. It should be understood, however, that automotive diesel progress was slower in the USA than in Britain because the economic need to obtain the utmost fuel economy was not so pressing in a country where cheap petrol was in ample supply.

The American Bosch equipment is virtually identical to the original Robert Bosch designs in Germany and with the British CAV components; minor differences occur in each, but the basic design and principles are the same. Several other pumps of the constant-stroke, variable-spill-control type were also made in America, the Adeco, Bendix-Scintilla, Demco, and Timken being prominent.

Another particularly interesting American pump was made by the Ex-Cell-O Corporation, Detroit. Although of the constant-stroke type with variable-spill-control, it departed distinctly from the standard 'Bosch type' in-line pumps because its cylinders were arranged in a circle around and parallel with the central driving shaft which reciprocated the plungers through the medium of a swashplate. A reciprocating feed pump integral with the unit delivered oil to a central reservoir and thence to the suction ports of the plungers. Control was effected by means of a sleeve on which triangular lands were machined. Axial movement of this sleeve caused the lands to cover the spill ports for a longer or shorter period as required, thus controlling the quantity of fuel forced through the delivery valves by the plungers.

A British pump, also of the axial type, was the Mono-Cam, at one time undergoing development by Simms. In this pump the plungers were actuated by a single-face cam mounted on the central driving shaft and regulation of

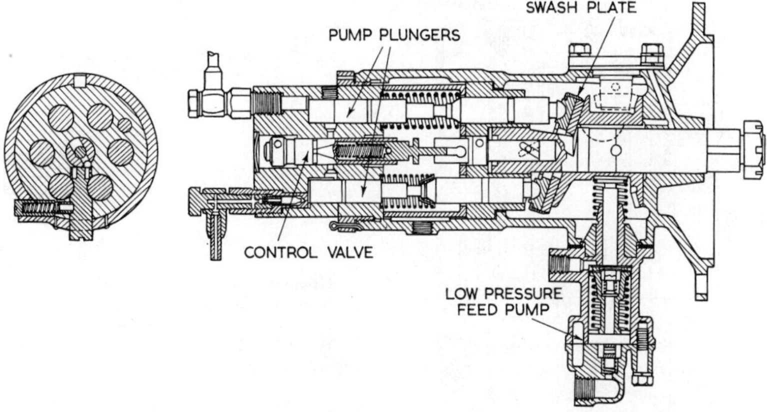

Flange-mounted Ex-Cell-O six-cylinder swash-plate injection pump

output was effected by a slidable sleeve which altered the relative position of the plungers and the barrels, so giving the normal effect of the spill method of control without partial rotation of the plungers.

A form of injection equipment widely used in America is the combined pump and injector in each cylinder head in place of the separate multi-element pump feeding individual injectors. The earliest of the modern diesels in the USA (the Cummins) was so equipped.

The original Cummins system employed a low-pressure metering pump with a rotary distribution valve so that each injector was supplied with the correct charge, the oil being delivered into the nozzle before the compression stroke began. The oil remained in the nozzle tip, being unable to flow through the fine spraying holes by gravity owing to their capillary effect. Discharge was effected by a plunger operated by a push rod and rocker, the oil being injected into the cylinder by the descent of the plunger to the full depth of the interior of the nozzle tip. During the compression stroke some air was forced into the nozzle so that a certain degree of aeration took place prior to injection.

The Cummins injection arrangement as described above was used continuously for upwards of 25 years until, in 1954, a new method of metering was announced as the PT (pressure-time) system. The same form of injector, directly operated by push rod and rocker, is mounted in each cylinder head, but metering is effected within this assembly and not, as previously, by the low-pressure pump on the crankcase which accordingly has been made both lighter and smaller.

In the PT system the supply pump feeds a fuel manifold to which all the injectors are connected but pressure in the line is varied according to engine speed and load, regulation being effected by a spring-loaded piston valve which co-operates with a series of spill or bypass holes. Rise of speed increases fuel pressure, while use of the accelerator delays release of the fuel pressure control valve. Fuel is admitted to the working zone of the injector through a calibrated aperture opened as the plunger lifts in the course of its normal cycle of operation. The duration of intake aperture opening is thus a function of engine speed, thus giving the time element (T) while the quantity of fuel

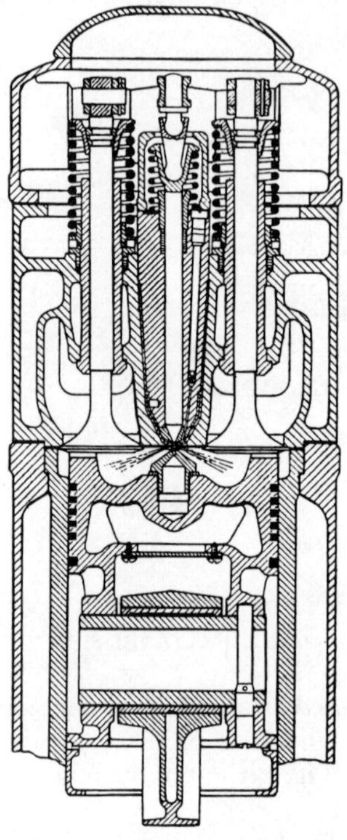

Cummins combined pump – injector. This injection system was used until the advent of the PT (pressure-time) system in 1954

admitted in unit time is a function of the supply pressure (P); hence the name PT system. The fuel pressure is varied with rising speed and load, both by the characteristics of the regulator valve and by the loading imposed thereon from the accelerator, with a centrifugal governor providing an overriding control to close down the fuel delivery under conditions of excessive rev/min. The only difference in the functioning of the injector from the earlier equipment is that the aeration of fuel in the nozzle tip no longer takes place.

An indication of the capabilities of the PT system is given by the recent introduction by Cummins of V6 and V8 diesels of 117·5 mm ($4\frac{5}{8}$ in) bore and 89 mm ($3\frac{1}{2}$ in) stroke, running at 3 300 rev/min and having a weight/power ratio as low as 3 kg/hp (6·4 lb/hp). Although the 'over-square' petrol engine has long been commonplace, particularly in America, it has been the practice to design automotive diesels with bore-to-stroke ratios of between 0·7:1 and

1 : 1. The lower inertia forces associated with a shorter stroke permit higher maximum speeds with a corresponding increase in power, but the higher speed inevitably requires an enhanced performance of the fuel injection equipment if combustion efficiency is to be maintained.

The widely used GMC two-stroke diesel also follows on similar lines but the usual constant stroke with variable-spill system is used, the pump and injector being formed in a single unit fitted into each cylinder head of the engine so that the pump plunger can be operated in the same way as the valves through a push rod and rocker from the engine camshaft. A feed pump

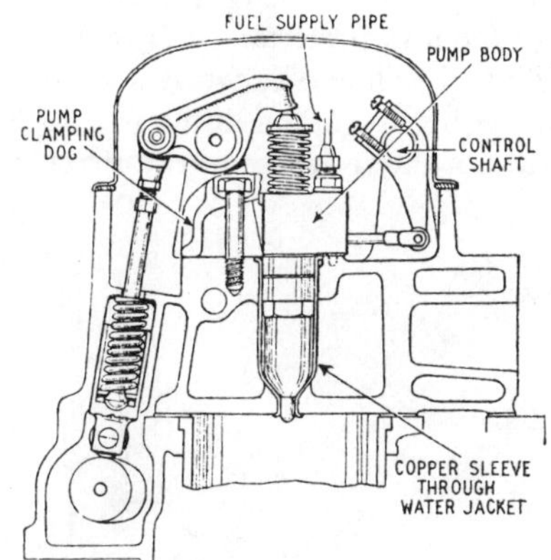

General arrangement of GMC combined pump–injector and its operation from the engine camshaft

delivers oil at low pressure to the injection pump and thereafter the process is the same as with other types of constant-stroke variable-spill equipment.

Since the injection pump is combined with the sprayer, the fuel oil is delivered directly from the one part to the other through drilled internal oilways and no external high-pressure piping or unions are involved; this, of course, is one of the major advantages claimed for the arrangement. On the other hand the control rod of each separate pump has to be coupled by external linkage to a control shaft from the governor. Thus, *phasing* of the pumps (arranging injection timing to be exactly the same for each cylinder), being a critical adjustment, is not so certainly effected when dealing with six individual pumps as with a compact multi-element pump which can be phased and calibrated at the bench on a pump test rig, after which it can be installed on the engine without risk of deranging the phasing by maladjustment of external control linkage.

Another possible objection to the combined pump and injector is that the whole unit must be removed for injector servicing, which is much more frequently required than attention to the pump. There is also the question of the desirability of having the pump subject to the heat of the cylinder head position, although this may be a matter open to different arguments according to whether the engine is operated in hot or in very cold climates.

The main factor affecting the choice of a combined pump and injector for each cylinder as against a multi-element unit pump with separate injectors appears to be the mechanical consideration of their respective adaptability to the type of engine involved. British and other European engine makers had the Bosch multi-element unit pump available from the beginning and engine design invariably was laid out to accommodate it; mounting dimensions of competitive alternative pumps quite naturally were designed to be interchangeable with the Bosch.

In America, the pioneer engine, the Cummins, preceded the wide availability of the unit pump in that country and the makers therefore designed a combined pump and injector assembly for each cylinder as integral component parts of their engine. Diesel development was rather slow in America for some years until the GMC two-stroke came into being and was soon produced in vast numbers. Once again a combined pump and injector was designed as an 'own make' component incorporated in each cylinder head. The enormous production of these engines warranted the design and manufacture of special equipment in preference to the use of a proprietary article.

Thus the choice between the various types of injection equipment appears to have been decided differently in Europe and in America as a result of the nature of the industry in the two continents, rather than on any question of

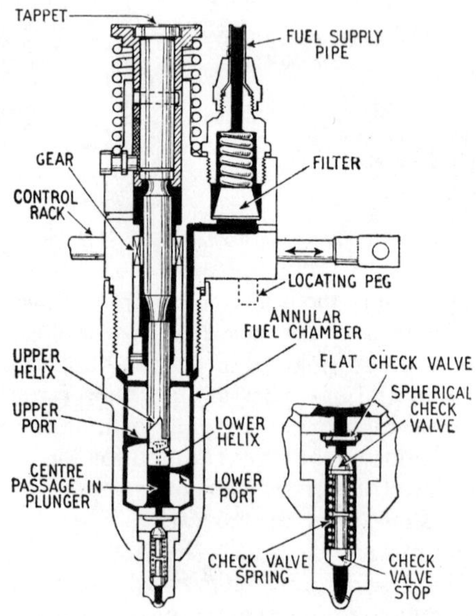

Section of GMC pump – injector unit

relative merit on technical grounds, although in this latter connection it may be observed that one of the limiting factors on valve dimensions is the amount of space taken up by even an ordinary injector. In the interests of volumetric efficiency, therefore, the more bulky combined pump–injector unit is open to objection.

Among the several American diesels based on German practice it is natural to find that the multi-element in-line pump has been used more freely, the equipment being either of the American Bosch type or one of those somewhat similar designs mentioned. However, the American Bosch Co. had concerned itself with the development of a high-pressure distributor system whereby the output from a single pump element could be directed in turn to the combustion chambers of a multi-cylinder engine. The pump camshaft had lobes corresponding to the number of engine cylinders and in its earliest form the single plunger was not only reciprocated by the cam lobes but was also rotated by means of a gear-driven guide.

The composite motion thus imparted to the plunger caused its outlet duct to register in turn with the pipe unions connected to the individual injectors; a suitable delivery valve was incorporated in the design. Output was regulated by a needle valve which provided a variable restriction in the fuel intake of the plunger from an integral low pressure feed pump. A governor of centrifugal, hydraulic, or pneumatic type could also be provided to act on the same valve.

Over the entire period of the development of the automotive diesel innumerable patents have been taken out in many countries relative to fuel pumps of the kind just described, although little progress was made in their practical application. The type is obviously attractive; the construction is relatively simple, it is compact, and it should be less costly as there are fewer parts which require high precision in their manufacture. Also the output must remain the same for all cylinders with only one pumping element; the need for careful balancing and calibration of individual elements is therefore eliminated.

Development in this direction has been greatly accelerated in recent years, and distributor pumps are a standard fitting on a large proportion of the British high-speed diesels now coming into service. Pumps of this type are manufactured in sizes suitable for engines of up to about 2 l. per cylinder.

An important feature of these pumps is their ability to operate at high speeds. With the exception of a pair of small reciprocating plungers, all the main components rotate and are thus freed from the inertia forces which are a serious consideration at high speeds in the spring-loaded cam-operated mechanism forming the basis of the in-line pump.

The American Bosch composite motion design did not mature on the lines indicated but in 1939 another distributor pump was developed by V. Roosa of Hartford, Conn., USA, and was put into production as the Roosa Master. The pumping element incorporated a pair of opposed plungers mounted radially in a body rotating within a cam ring. As the body rotates, the cams force the plungers towards each other during the delivery period, so ejecting the fuel between them. During the suction stroke, fuel is admitted between the plungers under slight pressure from an integral feed pump. An extension of the rotating body containing the plungers has a lateral distributor port which registers in turn with each of the outlet ports connected to the injectors by high-pressure pipes. Regulation of the output is by variable restriction of the fuel intake, which has the effect of controlling the outward travel of the plungers during the period when the inlet ports are in register.

The American Bosch Co. also developed a new distributor pump which, like the Roosa, had a pair of opposed plungers carried in a rotating element and actuated by their movement within an internal cam ring. In this pump the control of fuel output is brought about by the axial movement of a helically slotted sleeve which variably uncovers the low pressure supply ports, so allowing more or less spill back from the output to the input side. Governing is by means of a self-contained centrifugal mechanical arrangement consisting of an inverted conical cap which actuates the axially slidable control sleeve. The conical cap rests on a number of steel balls carried on the rotor and as these move outwardly under centrifugal force they raise the conical cap by their pressure against its conical undersurface.

In both the distributor pumps described a fuel feed pump is incorporated as an integral part of the design. In the Roosa arrangement this feed pump is a simple sliding-vane unit coaxial with the main rotating element. Not only does it maintain a continuous supply to the injecting plungers but, by the variable pressure of its delivery according to speed, it provides a means of hydraulic governing without the introduction of a separate mechanism. On the American Bosch design the feed pump is also on the rotating element and is of the vane type, the vanes being made of flexible material.

The basic principles of these pumps have set the pattern for most of the distributor-type pumps currently produced. For example, the DPA distributor

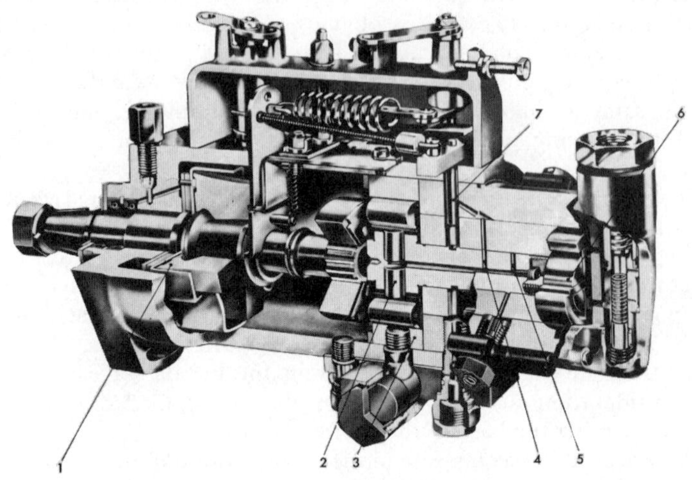

Construction of the CAV distributor-type fuel pump, type DPA, fitted with a centrifugal governor, Components include (1) centrifugal governor, (2) twin pump plungers, (3) cam ring, (4) suction port, (5) delivery port, (6) vane-type transfer pump, (7) metering valve

pump manufactured by CAV is similar in principle to the Roosa design. It is supplied with either centrifugal or hydraulic integral governors and it can be mounted either horizontally or vertically.

An interesting feature of distributor pumps is the comparative ease with which variable timing can be incorporated. In the CAV DPA type this is by fitting a spring-loaded piston which is actuated by the pressure developed by the vane-type transfer pump. As this pressure increases with speed, the piston compresses its spring and rotates the cam ring so that the timing of injection is advanced. Leakage of the oil permits the spring to return the piston, thus retarding the timing as speed is reduced.

As an indication of a possible trend in the future development of fuel injection equipment, there is the combined pump–injector unit designed by BICERI. This device is a hydraulically operated pump–injector unit, designed to fit into the space occupied by a standard injector. It contains a conventional nozzle unit together with a pump plunger actuated by a hydraulically powered piston, and a piston-operated feed valve. An engine-driven servo-pump unit supplies fuel at a constant pressure to a common supply line, and it also incorporates a distributor device for timing the actuation of each injector. Quantity control is by means of a control valve in the servo-pump unit which regulates the spillback from the feed valves. The main objective of this development was the problem of designing a satisfactory drive for the fuel pumps on highly supercharged marine and stationary engines where the large intermittent torques necessitate very massive shafts and gears. The BICERI pump–injector system has also been demonstrated as entirely suitable for use on small high-speed diesels, where the benefit of its more silent operation may be a factor in favour of its general adoption in the future.

Before closing this examination of the injection system some reference to the fuel feed from the tank to the injection pump must be made. A feature of the pump is that its plungers draw their charges of fuel from a common gallery in which the supply is maintained at a low but constant head, such as would be associated with a gravity tank mounted only a few inches above engine level. This, indeed, is the most simple and effective method of providing fuel supply, but it is no longer used on road vehicles because the general structural layout now requires the fuel tanks to be mounted low in the frame so that they are actually below engine level. In marine installations also, the fuel tankage is generally below the engine, so that only in the case of some portable industrial units and in stationary plant is natural gravity feed possible.

However, because the contents of the injection pump fuel gallery must be maintained at a constant head, any fuel lifting system provided must incorporate either a gravity stage as its final element or be provided with some kind of bypass valve which will ensure the stability of its low-pressure delivery, irrespective of fluctuations in rate of usage. When the automotive diesel first came into use the most common fuel lifting equipment was the Autovac, a 'suction' device actuated from the exhauster which was driven by the engine for the operation of the vacuum brakes. In the Autovac system suction was applied to the fuel line through a float-controlled valve so that a container mounted above the engine level was filled from the low tank. The container was connected directly to the injection pump through the medium of suitable filters; thus the fuel gallery supply was maintained under the gravity head of the Autovac container, which was automatically replenished by intermittent application of vacuum to the tank suction line under the control of a float valve. Although the Autovac system is now rarely fitted to new vehicles it continues to function well on many buses still running.

The present fuel lift equipment is of the mechanical pump type, either plunger or diaphragm, actuated from the injection pump camshaft and, indeed, generally mounted therewith as an integral unit. The reciprocating fuel lift pump is of simple construction having a constant-stroke plunger actuated by an eccentric. This type of pump is extensively used, but it presents certain problems, the most serious being that it can build up pressure greatly in excess of requirements, especially under operating conditions when the injection pump is only disposing of small quantities of fuel. To meet this the filter in the delivery connection must be provided with a sensitive relief valve connected by an overflow pipe to the main fuel tank. An alternative method is to discharge from the lift pump into an air-vented service tank above engine level, from which the injection pump is fed by gravity. An overflow from the service tank to the main tank is also essential.

To reduce this somewhat extensive 'plumbing', the alternative type of diaphragm lift pump is often used. This is similar to the engine-driven petrol pump used on the majority of modern motor cars, but it has a diaphragm of

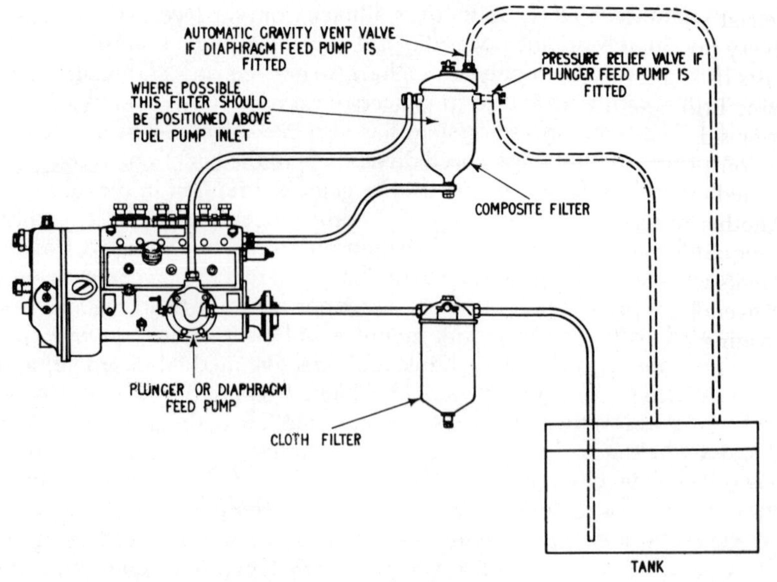

Typical arrangement of fuel feed by lift pump from a low-level tank. There are two filters in series, one on the suction and the other on the delivery side of the lift pump

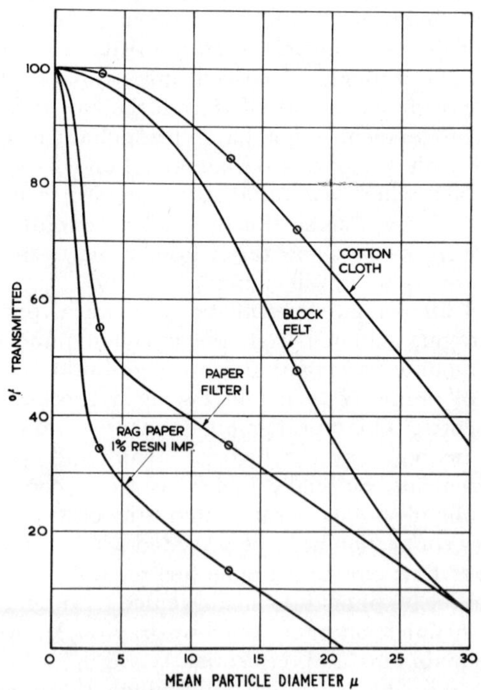

Comparative efficiency of various recognised filtering media as tested by CAV

material which is resistant to damage by fuel oil. This pump maintains a delivery pressure of about 0·22 kgf/cm^2 (3 lbf/in^2) (or less) and automatically adjusts its output to requirements. Thus, within limits, if the injection pump is using little fuel the diaphragm pump ceases to lift until the pressure on the output side falls. The diaphragm pump may be mounted directly on the injection pump or it may be actuated from the engine camshaft, when it is mounted on the side of the crankcase.

Another most important matter connected with fuel feed is the provision of adequate filtration against the passage of abrasive particles to the pump elements with the risk of damage to their highly finished surfaces. The latest pumps embody extremely fine filters, but these must be regarded as a final and overriding protection which supplements the usual two external filters in the

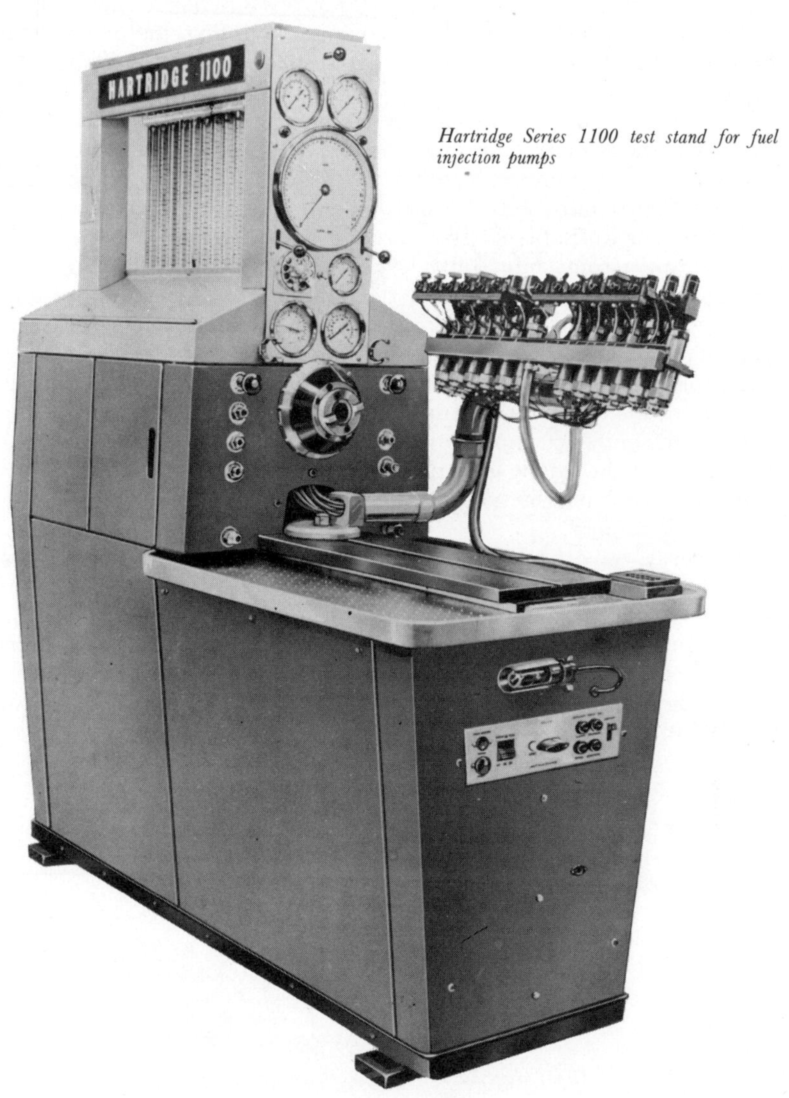

Hartridge Series 1100 test stand for fuel injection pumps

line between the fuel tank and the injection pump. These filters generally consist of cloth, felt or a special kind of paper, or possibly a combination of two of these materials. They may be located in various positions, but a common arrangement is to have one between the tank and the lift pump and the second, and finer one, between the lift pump and the injection pump. This latter filter is usually placed so that it will receive engine heat to ensure free flow under cold conditions; it is also provided with a vent or relief valve piped back to the tank.

A class of fuel filter that has rapidly attained great popularity is the paper filter with easily replaceable 'throwaway' element. The filtering medium is a specially prepared paper which is folded in radial pleats in order to present the maximum surface area to the fluid passing through it. The degree of filtration in terms of particle size is measurably superior to that of the felt or cloth materials previously used. To facilitate handling, the paper element is supplied in a metal can which is directly substituted for the earlier types in any standard filter bowl.

It is obvious that the satisfactory running of a high-speed diesel engine depends perhaps more than anything else on the correct functioning of the fuel injection equipment; periodic reconditioning and calibration are essential, and require considerable skill. An extensive range of sophisticated devices is now available for the servicing of pumps and injectors, and these devices range from comparatively simple test kits for checking and fault finding on the engine, to elaborate test stands which are fully instrumented with elegant electronic devices for the complete testing and calibration of all types of fuel injection equipment.

Complementary devices cover the stripping, reconditioning, adjustment, and calibration of pumps and injectors. Various jigs, gauges, optical devices, lapping machines, cleaning equipment, and many other special tools are required. It is also vitally important that the servicing of fuel injection equipment is carried out in specially designed workshops under conditions of absolute cleanliness.

Obviously, only the operators of large fleets are likely to be able to afford the considerable outlay involved in providing adequate facilities, and it has become common practice for much of this work to be carried out by firms specialising in the servicing of fuel injection equipment. Manufacturers who have developed a comprehensive range of servicing, testing, and calibrating equipment include Leslie Hartridge Ltd., of Buckingham, and the Merlin Engineering Co. Ltd., of Halifax.

6
Combustion chamber design

In the early stages of the development of the modern high-speed diesel the varieties of combustion chamber layout were so numerous as to confuse the ordinary user. Each design was put forward with great claims for ease of starting, improved combustion, greater economy, smoother running, higher speed, freedom from diesel knock, and so on. Not all these claims proved to be justified. A recapitulation of the ideas associated with this period serves to explain an interesting phase, even though certain of the once promising designs had already become historical rejects in the space of less than 20 years. Consideration of these early efforts, however, reveals something of the problems which had to be dealt with and there is no doubt that the intensive work carried out in the early 1930s produced that clear understanding of combustion phenomena which resulted in the stabilisation of design around two or three basic types by 1939.

In this country the accent was upon the thermal efficiency reflected in favourable specific consumption, and in this respect the British automotive diesel established a world standard, primarily because of its concentration upon the direct-injection (or open) combustion chamber cylinder head design, coupled, of course, with the recognition of the influence of fuel-injection characteristics in attaining the same end.

On the Continent the line of least resistance was followed, the aim being directed towards securing a steady and progressive burning of the fuel by a system of pre-combustion, the charge being ignited in a partly separated chamber from which the more or less controlled expansion then passed to the working cylinder. There were very reasonable and logical grounds for pursuing this system, even though later experiment and development indicated that there was a degree of error in the basis of the reasoning. The antechamber type of engine certainly has been very highly developed, and most engines employing it, or some modification thereof, are either of Continental origin or are based on Continental patents.

In the antechamber combustion head, when the piston rises to the limits of the cylinder head, air is compressed into a chamber connected to the cylinder proper by a passage, or passages. This device, termed the atomiser, is of special form, in the nature of a cap through which a number of fine holes are drilled. By suitable design the chamber is kept at a fairly high temperature and some degree of turbulence is imparted to the air in its passage through the connecting holes. Fuel is injected into the chamber, and ignition of a part of the fuel–air mixture gives rise to rapid expansion, which drives about three-quarters of the still unburnt fuel particles through the holes in the atomiser, giving a finely divided spray into the cylinder proper, where combustion continues.

Modifications of the system lie chiefly in the disposition of the antechamber and the general shape of the cylinder head, but the main principle is the same in all engines which are of the true pre-combustion type.

Attractive as the antechamber arrangement appeared to be, it cannot be said to have been an unqualified success on small engines for transport and

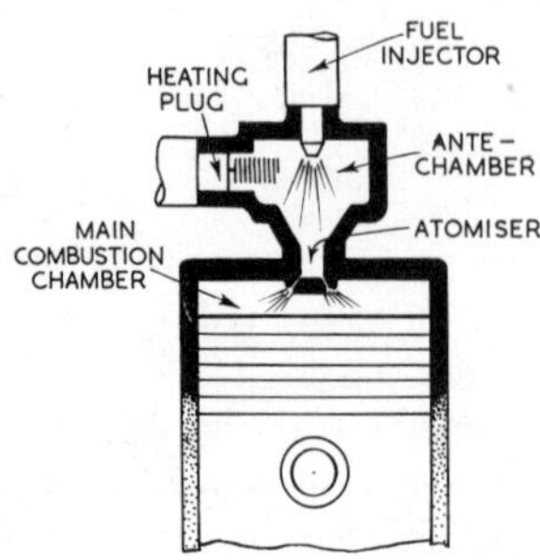

(Left) *Antechamber or pre-combustion head arrangement*
(Below) *The Büssing antechamber with three divergent holes in the tip co-operating with 'broad arrow' depressions in the piston crown. The nozzle tip is shown partly sectioned. Büssing also make direct-injection engines with toroidal combustion chambers*

similar applications, where wide variations of load and speed prevail. A high compression ratio (up to 21 : 1) is called for, and almost invariably powerful electric starters and heater plugs have to be used.

In the confined space of the antechamber some of the fuel at any rate must come into contact with the hot walls of the atomiser, while the shape and size of the connecting holes must necessarily be planned with more regard to their primary function of atomising than of imparting turbulence to the compressed air; avoidance of partially burnt fuel and fuel carbonised on the hot walls may, in consequence, be difficult over a wide range of injected quantities of oil. Moreover, the volume of the antechamber is unswept by the incoming air, and although the products of combustion remaining after the exhaust valve closes may be drawn out during the induction stroke, they dilute the intake air. Finally, the work required to force the air in and out of the antechamber is a 'pumping loss' and therefore affects the efficiency of the engine.

In popular terms, however, the combustion process in the antechamber engine was planned to give a more sustained push to the piston than would be the case if the burning of the charge took place in the open cylinder, for it was assumed, because of the phenomenon of diesel knock, that ignition in this type of engine was much more rapid than in the case of the petrol engine. This was true only up to a point and it was some time before a distinction was drawn between ignition and combustion when the latter term was considered as a complete process.

In the compression ignition engine there is a time lag between the start of fuel injection and the initiation of combustion, which is one of the main causes of diesel knock. As the true state of affairs was not recognised at the time, an endeavour was made to control the combustion process by the device of the pre-combustion chamber, and it was only after some years that it was possible

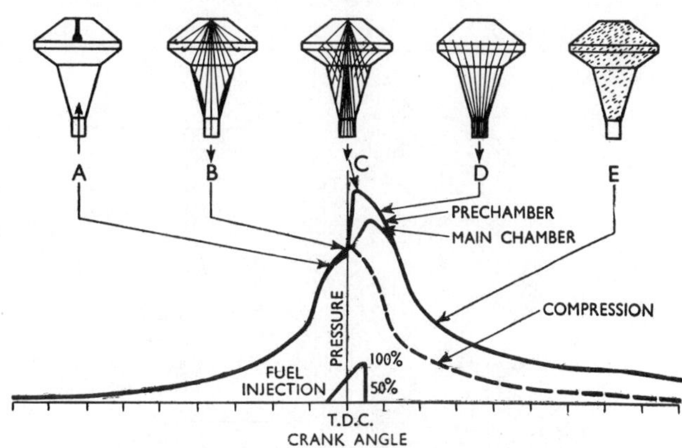

Phases of combustion in Caterpillar antechamber:

A. *Injection starts; air flowing into chamber*
B. *Ignition; fuel build-up on throat walls*
C. *High pressure; fuel-air mixture flows from chamber*
D. *Injection ends*
E. *Combustion ends; residual products*

to obtain controlled and complete combustion in an open chamber by the more intimate and homogeneous mixture of fuel and air, made possible by controlled turbulence and by the development of improved fuel injection equipment, together with the production of fuels of improved ignition quality.

One of the most difficult problems confronting the designer of high-speed diesel engines has always been the avoidance of smoke in the exhaust at all loads and speeds. This can be achieved only by the perfect matching of the fuel spray pattern to the shape of the combustion chamber and the turbulence therein, so that a homogeneous distribution of uniformly small droplets of fuel throughout the entire volume of the combustion chamber is achieved. Any small pockets where the quantity of fuel is in excess of the air immediately available for its complete combustion will lead to the production of smoke. A very small quantity of incompletely burned fuel can produce considerable smoke in the exhaust, even when the specific fuel consumption of the engine is still quite low.

From the pure antechamber or pre-combustion chamber, the next development was the air cell. In this type of cylinder there is a separate chamber connected to the cylinder head by a shaped passage of venturi form, that is, with a narrow neck expanding on both sides of the constricted part. In this type of engine the rising piston compresses the air in the head and drives it into the air cell. Usually the fuel jet is arranged to spray the oil into the communicating venturi passage. Ignition takes place, however, before the fuel completely enters the air cell, and, as the piston descends, the air from the cell returns to the cylinder, a 'blowlamp' effect being reproduced at the mouth of the venturi, still further mixing the fuel and air as the combustion continues. The air cell type of engine has shown good results up to a certain point, and even if it does not achieve its object of burning the fuel completely over the entire range of speed and load of which it is capable, it has certainly shown a capacity for comparatively high speeds.

Such shortcomings as it has seem to be traceable to the fact that a fairly large pocket remains unswept by the incoming air charge and a big percentage

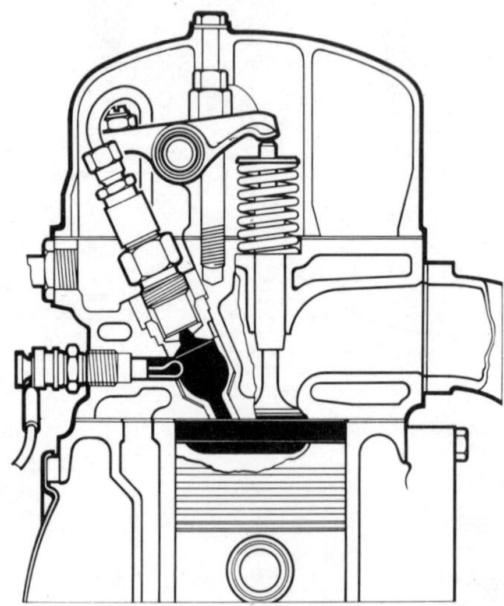

In a later form of Mercedes (Daimler-Benz) engine the pepper pot type of antechamber gave way to this single-orifice air cell

volume of the compression space is filled with hot inert gas at the commencement of compression; thus the higher compression ratio employed has a somewhat fictitious value. The relatively large wall area with which the compressed air is in contact also accounts for an appreciable heat loss.

The spraying problem in the air cell type is also difficult. In the parent Acro system (of German origin) the spray is directed into the throat of the venturi and a very narrow angle spray is needed. Variations of the arrangement aim at passing the spray through the moving air stream at a slight angle, but, according to the constricted shape of the venturi passage and the disposition of the water-jacketed parts of the head in the vicinity, either condensation or carbonisation of the fuel may result during the delay period following ignition.

In common with the antechamber type, the air cell engine requires a high compression ratio, and, often, heater plugs for starting. But, in its realisation of the value of controlled turbulence as a medium for the thorough mixture of fuel and air, it marked a distinct advance.

Several variations in the shape of the air cell have been evolved, from the lobe to the spherical, with offset or symmetrical connecting passages and with wide conceptions of the best type of venturi, all being aimed towards the attainment of perfect combustion.

Of the type of air cell engine in which the cell is located in the piston it is rather difficult to form any opinion other than that this was a passing phase planned chiefly with a view to avoiding costly production changes in existing types of power unit. When higher speeds are being sought it is desirable to reduce piston weight as much as possible, whereas the air cell increases it. In addition it must be a difficult item to retain at a uniform temperature, and local hot spots are to be avoided in piston design.

Moreover, there seems no logical reason why the air cell should not be in the cylinder, since the desired effect—especially when it is used merely to prolong the combustion period—should equally be attained whatever the location of the cell in the combustion space.

Reference should also be made to the Lanova air cell combustion head developed by the German engineer Franz Lang, the originator of the Acro

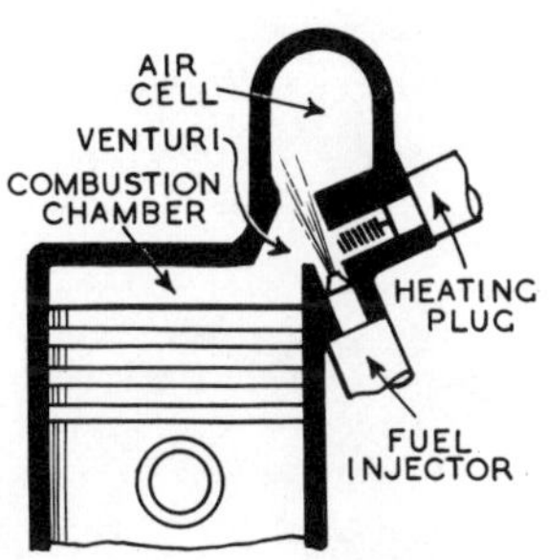

Air cell (Acro system) cylinder head

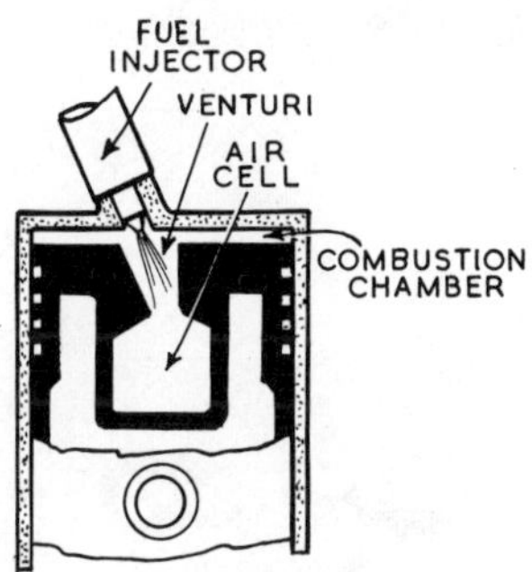

Acro air cell piston

air storage chamber. It is designed to afford controlled turbulence or air flow with an exceptionally low proportion of excess air. The original combustion chamber had the form of a figure 8. In line, opposed, and directed into the waist of the 8, are the fuel sprayer and a cone-shaped air storage cell having a funnel-shaped opening. This opening communicates with the main chamber on one side and on the other leads into a second air cell through a passage that can be opened or closed by a plunger so that the capacity of the combustion space may be reduced; when it is reduced the compression pressure is raised and so facilitates starting. After starting, the second air cell can be brought into operation. Upon injection, the fuel enters the first cell and part of it also the second. In these cells it ignites, and the pressure in them immediately rises so that the burning charge is blown out into the main combustion chamber. But as this process is throttled by the gas having to pass through the relatively small diameter apertures, the pressure rise in the main combustion space is gradual. Combustion is, moreover, considerably improved by the burning mixture of gas and fuel whirling round in the two loops of the 8, so that the air and fuel are intimately mixed. The form of combustion chamber used and the location of the injection nozzle permit the use of large valves, resulting in high volumetric efficiency.

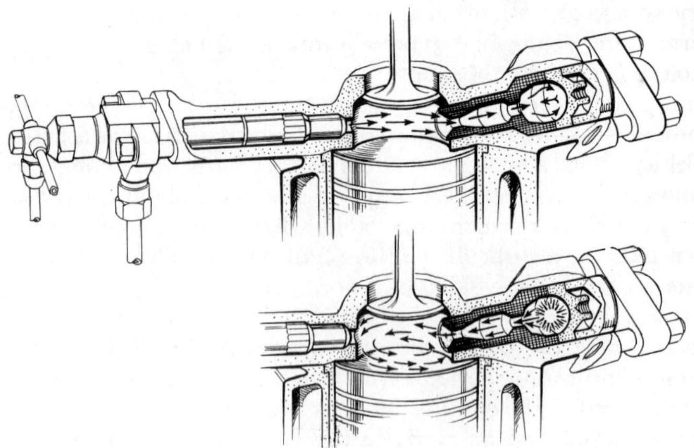

In America the Lanova combustion system has been developed as a single-lobe chamber with injector and 'energy cell' placed tangentially to promote swirl

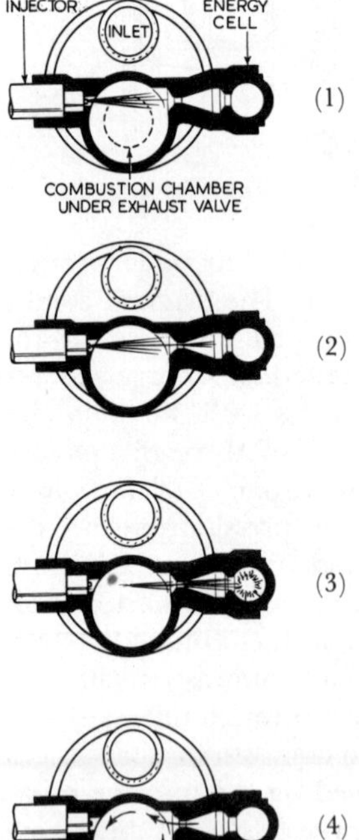

Phases of combustion in the Lanova system:

(1) *Injection starts in the single-lobe head*
(2) *Early fuel penetrates to the energy cell*
(3) *Injection ends, ignition starts in energy cell, pressure rise spits back air and fuel through venturi into combustion chamber*
(4) *Swirling combustion – the final stage in the cycle*

The Lanova system has found more favour in America than in Germany, its country of origin. It provides smoother performance than is associated with other German air cell designs and permits reasonably high speeds, but the specific consumption is not particularly favourable, judged by British standards. Incidentally, the manually operated closure of the second air cell to raise the compression ratio for starting now appears to have been discarded, while in USA the combustion chamber has been modified as a single lobe. In Britain both Dennis and Meadows at one period built engines with Lanova combustion chambers, but neither continued to use the system.

Another air cell system with controlled tubulence was a cylinder head design patented by the Omo AG of Zurich. It was employed for the Vomag and Hansa Lloyd engines in Germany, and the Rochet Schneider in France; in Britain it was used experimentally by Tangye, Ltd., while the American Hercules engines are based on it although with modifications.

The principal feature was the special shape of combustion chamber, spherical and located at one side of the cylinder head with a passage communicating with the cylinder. Of thin metal, the chamber was detachable, and being surrounded by an air space took up heat quickly on starting. The combination of the chamber and communicating passage produced a condition of controlled turbulence on the compression stroke, during the greater

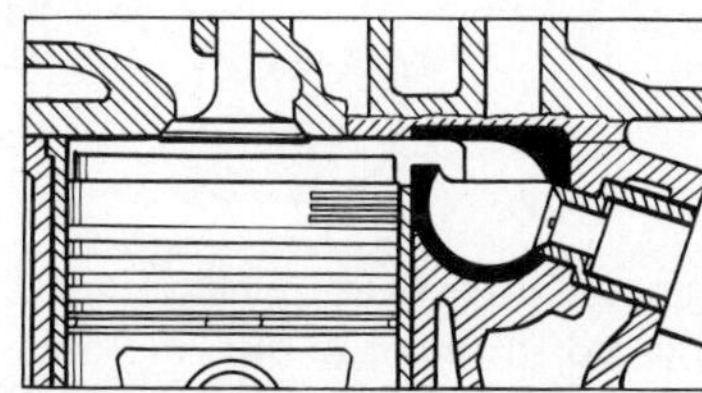

Lateral swirl chamber of the Omo type used in American Hercules engines

part of which the air entered the chamber freely, producing a strong swirl into which the fuel was injected across the stream through a pintle-type nozzle, so that the fuel and air were intimately mixed; the combustion of the fuel occurred mainly in the chamber. Starting was assisted by an electric heating coil, inserted horizontally, but the spherical chamber very quickly reached the working temperature.

It should be recognised that pre-combustion and air cell engines in general are not very critical as to fuel quality and this has had much to do with their wide acceptance in both Europe and USA, where the rigid standards of highly refined diesel oil as used in Britain have not always been insisted upon.

A great advance in combustion chamber design was made in this country by H. R. (Sir Harry) Ricardo, the underlying principle being the projection of the fuel across a rapidly rotating mass of air in the combustion chamber in such a manner that the air has to find the fuel rather than the fuel the air. In the Ricardo Comet head this air rotation is produced during the compression stroke by forcing the air through a relatively large tangential passage into a partly separated spherical air cell into which the fuel is injected.

The Comet head gave greater flexibility and freedom from smoke than any previous air cell design and it was adopted by many prominent British engine makers, notably AEC, Crossley, Dorman and Thornycroft, while overseas makers who adopted it were Berliet, Citroën and Renault in France, Fiat in Italy, Brossel in Belgium, and Waukesha in America.

A later design of Ricardo Comet head, known as the Mark III type, showed further improvement and was generally adopted by those makers who had used the previous type. The principal alteration is that two hemispherical depressions were provided in the piston crown on the same side of the cylinder as the spherical combustion chamber, which was of smaller capacity, while the injector was placed horizontally instead of being inclined at 45° from the vertical

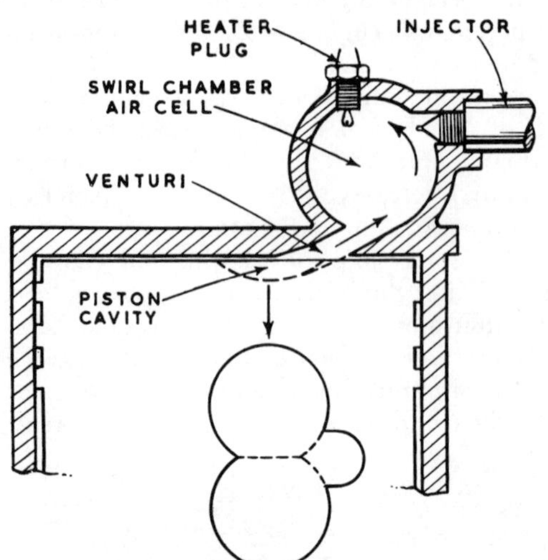

Ricardo Comet indirect combustion system. Swirl chamber combustion utilises rotary motion in the cell produced by the air being forced through the tangential venturi by the rising pistons

as with the older head. The tangential passage from the combustion chamber was placed so that the burning and expanding air-fuel mixture was directed down the sides of the depressions. With this design not only was the fuel consumption improved, but the running was smoother, especially at idling speeds, and there was a slight increase in power. Ricardo's contributions to the knowledge and understanding of the essentials of diesel combustion distinguish him among research engineers, particularly in respect of his recognition of the importance of turbulence.

The purely British high-speed diesel movement, however, was initiated by a direct-injection engine in 1929, and that principle has been adopted increasingly, even by the most persistent supporters of the air cell combustion chamber. As a result, in the exacting field of road transport work the direct-injection type is now almost universal in British automotive engines. It does not follow that it is the best, but rather that, on balance, if offers the most practical solution to many of the difficulties that have been associated with the application of diesel engines to transport vehicles. Even though other systems are used with success in various fields, there is now a definite trend to accept the direct-injection automotive type in marine, railcar, portable and light industrial applications.

From what has been said of pre-combustion and air cell designs it will be appreciated that the requirements in the engine are (1) high enough initial compression to produce the necessary heat for self-ignition and (2) a form of combustion chamber conductive to effective mixture of fuel and air. How are these met by the open cylinder?

In an open-cylinder engine, compression space can be reduced to give any desired compression ratio, and the only limiting factor is clearance between the valves and the piston top. Actually, in a symmetrical cylinder head a compression ratio of about 12 or 13:1 (about 21·3 kgf/cm^2 or 300 lbf/in^2) provides sufficient heating of the air for ignition, and even for starting from cold, except in the case of small engines. But in its simplest form this type of head does not give the turbulence desirable for thorough mixing of the air and fuel. However, it has the advantage of less dilution of the incoming air charge by residual exhaust gas, and it presents less surface area to the charge so that heat loss to the walls of the combustion chamber is reduced. In this respect, a spherical combustion chamber would be nearer to the ideal, because a sphere has the lowest possible surface-to-volume ratio, but there are practical difficulties with this form. For example, it would be difficult to arrange the valves in a hemispherical cylinder head, and, by way of compromise, many engines have combustion chambers of hemispherical or shallow bowl form, an example being the Gardner which favours a hemispherical bowl in the piston crown in association with masked inlet valves.

Thus we have the direct-injection engine, highly efficient from most points of view, capable of certain cold starting, but in its cruder forms liable to roughness and possibly to incomplete combustion, while the antechamber and air

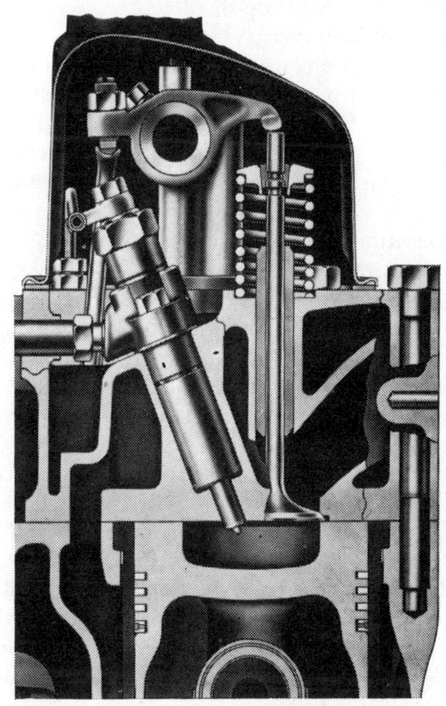

The original Ford combustion chamber; this was a simple open cavity in the piston, with an inclined four-hole injector. Ford now favour a toroidal combustion chamber and the injector is now mounted outside the rocker box cover

cell engines are seldom true cold-starters, are somewhat less efficient, but tend to softer characteristics in running.

All design, however, is a matter of compromise, and the direct-injection engine has been intensively developed. By the use of suitable valve ports, by applying masking devices to the heads of the inlet valves, and by so designing the shape of the piston crown as to displace the air violently as maximum compression is reached, a sufficient degree of turbulence can be obtained (coupled with the most suitable type and disposition of the spraying nozzle) to attain a satisfactory degree of mixture of fuel and air. A distinction must here be made between intake turbulence initiated at the inlet port and providing a rotational air swirl about the cylinder axis, and 'squish' turbulence

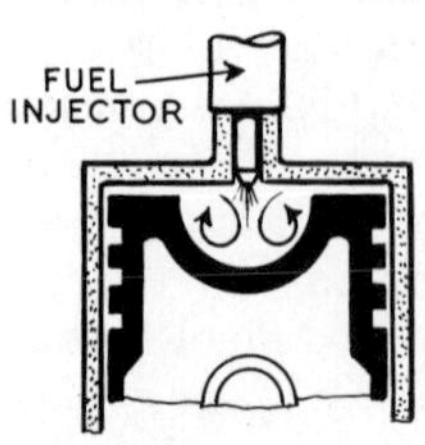

Showing use of the piston crown contour to promote 'squish' turbulence

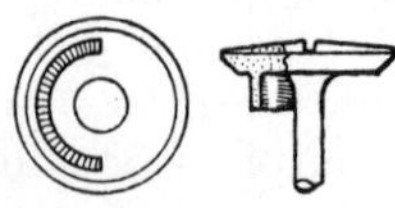

Masked inlet valve to direct swirl of entering air

caused by the shape of the piston crown in co-operation with the cylinder head. Squish turbulence is a horizontally projected inwards radial movement which occurs at top dead centre. The two forms of turbulent air motion can be combined in varying degree.

It is significant that in recent marine installations and on road transport vehicles, where progress has been most rapidly and intensively made during the past few years, direct-injection engines are now accepted as the most popular type. The efficiency of the direct-injection engine needs no proof. It has been developed very rapidly by several British makers to a state of high performance, and the practical results of the type both in road and marine operation are now acknowledged and are indeed a credit to the engine manufacturers.

In particular the direct-injection system provides easy and certain cold starting and is outstanding in specific consumption.

Some of the most important considerations in the open cylinder engine are bound up with fuel injection, while the pressure of injection, the location, penetration and direction of the spray have all had to receive infinite care and attention. Atomisation and turbulence, however, have reached a stage in which very satisfactory combustion is secured over a wide range, and the power output figures for a given fuel consumption are generally better than in the case of any pre-combustion or air cell type. Furthermore, the work on the direct-injection engine has shown perhaps more conclusively than in any other direction that the reduction of diesel knock is primarily a matter of proper combustion.

A milestone in the history of the automotive diesel was set up by the Saurer Co. of Switzerland when in 1934 it announced its dual-turbulence system. The two inlet valves of its four-valve head were masked to promote rotary air swirl about the cylinder axis during the induction stroke, while a heart-shaped cavity in the piston crown received squish at the end of the compression stroke. Owing to the persistence of the axial rotation, the two turbulences were combined during the fuel injection phase so that the four sprays from the multi-hole nozzle were directed across an air stream moving both axially and radially, the resultant air movement being spirally around the piston cavity.

This motion is described as toroidal, and this name has been subsequently applied to the cavity as a type designation.

The dual-turbulence system was introduced to Britain on Armstrong-Saurer engines; after their withdrawal from the British market, the toroidal-cavity combustion chamber (with only one inlet valve, however) was adopted by many of the foremost British engine makers and in particular by those with high output facilities. Thus a very large proportion of British automotive diesels now embody variations on this form of direct-injection combustion chamber, ranging from quite shallow bowls with a central displacement dome to quite deep toroidal cavities with re-entrant sides, as used in the original Saurer version. It is significant that many manufacturers have now given up the use of masked inlet valves which have the disadvantages of being more expensive, of imposing a restriction on the flow of air through the valve, and of involving maintenance difficulties. It has been found that careful shaping of an inclined inlet port can impart adequate swirl to the incoming air without the use of a masked valve.

The characteristics of a direct-injection system are well illustrated by the comparative performance curves of two similar AEC engines, one with this type of combustion chamber and the other with an air cell combustion system. These show clearly that the air cell engine allows for higher rev/min and a

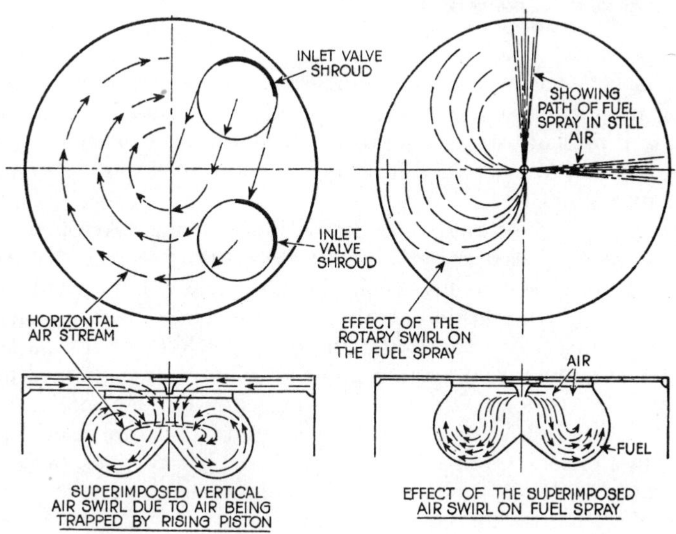

Air paths in AEC direct-injection engines

corresponding increase of power, whereas the direct-injection engine, although limited in speed and power, shows a much better torque curve which is substantially flat over the working speed range; the consumption figure is noticeably better and it is this factor, more than any other, that has caused its general adoption. In turn, the stress on fuel economy is the direct result of the relatively high cost of fuel in Britain arising from the heavy tax imposed on imported hydrocarbon oils used in road vehicles.

Before the present trend towards the development of small engines began, it seemed likely that the air cell combustion chamber might have a diminishing place in the British scene. Direct-injection arrangements were associated with practically all engines of not less than 114 mm ($4\frac{1}{2}$ in) bore, of $1\frac{1}{2}$ l. per cylinder, or $8\frac{1}{2}$ l., for a six-cylinder engine. However, units of this type were scaled down to about $6\frac{1}{2}$ l., while several four-cylinder engines of from 4 to $4\frac{1}{2}$ l. also came into being with bore sizes of about 101·4 mm (4 in) or rather less, the swept volume of individual cylinders being about one litre.

Below these sizes, that is, in the case of four-cylinder engines of $2\frac{1}{2}$ l. or less and of about 76 mm (3 in) bore the air cell chamber continues to hold its

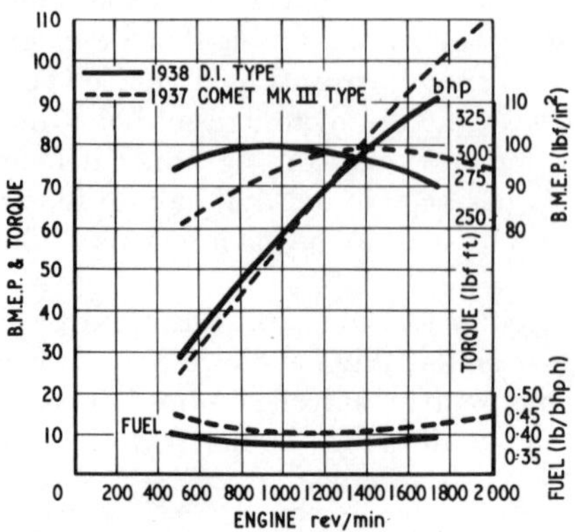

Comparison of performances of AEC indirect and direct-injection engines

own and, in fact, has even greater numerical representation because of the increasing number of small engines. The Ricardo Mark III chamber, for example, is at present used on the BLMC (Austin and Morris) 2·2 l. engine, while the Ricardo Mark V modification is incorporated in the BLMC 1·5 l. and 950 cm^3 engines.

The range of Perkins engines embodying the famous Aeroflow air cell combustion system has been well known since before World War II. The system is interesting as it claims to combine the merits of both the air cell and the direct injection combustion combustion chambers. The two-hole nozzle is arranged to direct one spray into the air cell while the second spray is directed towards the throat leading into the cylinder. It is claimed that the second spray meets hotter air, thus assisting early ignition and easier cold starting. It is interesting to observe, however, that in providing for small engines with a capacity from 500 cm^3 to 750 cm^3 per cylinder, both Perkins and Ricardo chamber designs were modified on somewhat similar lines. The air cell on the 1·6 l. Perkins engine and the Ricardo Mark V type both tend to a more pronounced beehive shape with a much shorter tangential throat passage.

In the Perkins design the customary two-hole sprayer used in the Aeroflow design has been replaced by a single-hole pintle-type nozzle positioned to inject in opposition to the air flow, while in the Ricardo design a 'Pintaux' nozzle is used. It is a feature of both that the bottom part of the air cell containing the throat is a separate ring of heat-resistant steel inserted in the mouth of the cavity machined in the cylinder head.

The current trend, and a major part of Perkins production output, is the use of direct-injection engines, and more recently has been added the development of fully machined inlet porting. This gives consistent output without

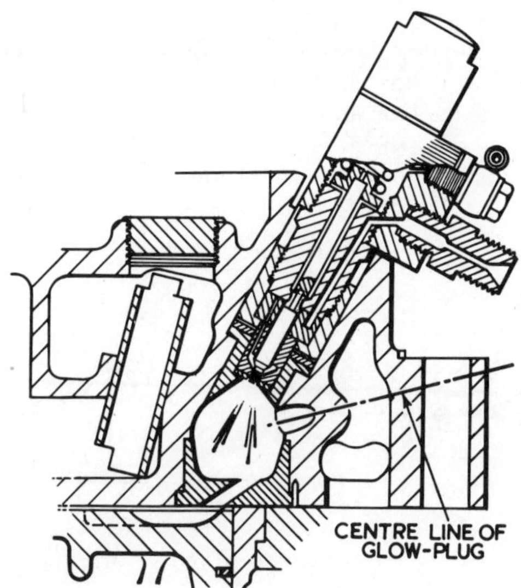

Ricardo indirect (Mark V) combustion chamber with CAV Ricardo-Pintaux injector. This differs considerably from the Mark III type used on many engines

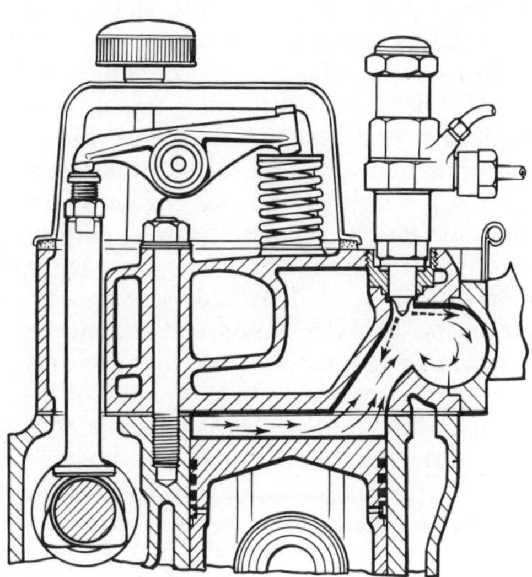

Perkins Aeroflow air cell combustion system with two-hole injector

deviations so often found with cast porting, and there is no need for valve deflectors.

It is interesting to note that the air cell and antechamber types of combustion chamber do not appear to have had much attraction for the designers of high-speed two-stroke diesels. The uniflow scavenge system, with a belt of inlet ports completely encircling the cylinder, can produce ample swirl by means of the tangential arrangement of the ports, and additional turbulence can be

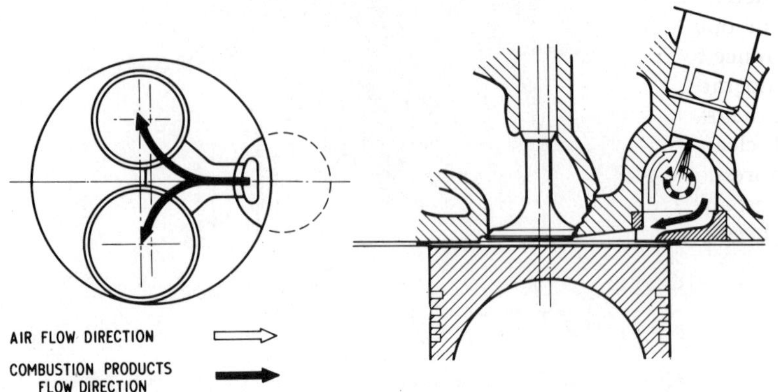

The Perkins H-type indirect-injection combustion chamber used in higher speed examples of the range, running at 3 600 and 4 000 rev/min.

obtained from the squish at the end of the compression stroke, so fairly simple combustion chambers are a feature of these engines. The Foden two-stroke design favours a fairly shallow toroidal bowl in the piston crown into which the fuel is sprayed tangentially by an inclined single-hole injector mounted near one side of the cylinder, while the GMC two-stroke engine makes use of a somewhat shallower toroidal bowl with a multi-hole injector mounted centrally in the cylinder head (see illustration on page 175).

On the other hand, the combustion chambers used in opposed-piston two-stroke engines almost invariably take the form of a flattened sphere formed by shallow bowl-shaped depressions in the piston crowns. In this case, there is the problem of obtaining satisfactory penetration and distribution of the fuel from the fuel injectors which spray radially (or tangentially) into the combustion chamber. In the case of the Rootes opposed-piston two-stroke engine, one single-hole injector per cylinder is used, but in some of the larger opposed-piston engines now coming into service, two diametrically opposed injectors have been fitted.

A feature of the development of two-stroke diesel engines in the late 1930s was the introduction of the Kadenacy system which received much attention at that time. In this system, the pressure waves in the exhaust passages are used to assist with the induction of scavenge air into the cylinder. The design of the exhaust manifold, the timing (and the rate of opening) of the exhaust valves or ports, and the timing of the opening and closing of the inlet ports are all arranged in such a manner that the momentum of the column of exhaust gas produces a momentary depression in the cylinder. This depression can be used to assist the inflow of scavenge air, thus lessening the power required to drive the scavenge blower. Experimentally, it can be demonstrated that a uniflow two-stroke diesel engine can run without a scavenge blower, once it has attained its running speed with the assistance of an auxiliary scavenging device such as an air hose directed into the inlet ports. In practice, most automotive two-stroke diesels take advantage of the Kadenacy principle, but invariably they are fitted with engine-driven scavenge blowers to ensure adequate flexibility. The exhaust pulse phenomenon can also be applied to a four-stroke engine but the benefit is less noticeable as thorough scavenging is inherently easier to achieve in the case of the four-stroke cycle.

The operating cycle of the uniflow two-stroke diesel can be described by reference to the schematic diagram on p. 75 showing the cylinder layout and the operational cycle of the Chrysler opposed-piston two-stroke engine. In the first diagram scavenging has been completed, the cylinder has been charged with clean air and the compression stroke is starting. In the second diagram compression has been completed, as the pistons have reached the point of their nearest approach, and fuel is injected. In the third diagram the working stroke has been completed and at this point (about 70 crankshaft deg. before outer

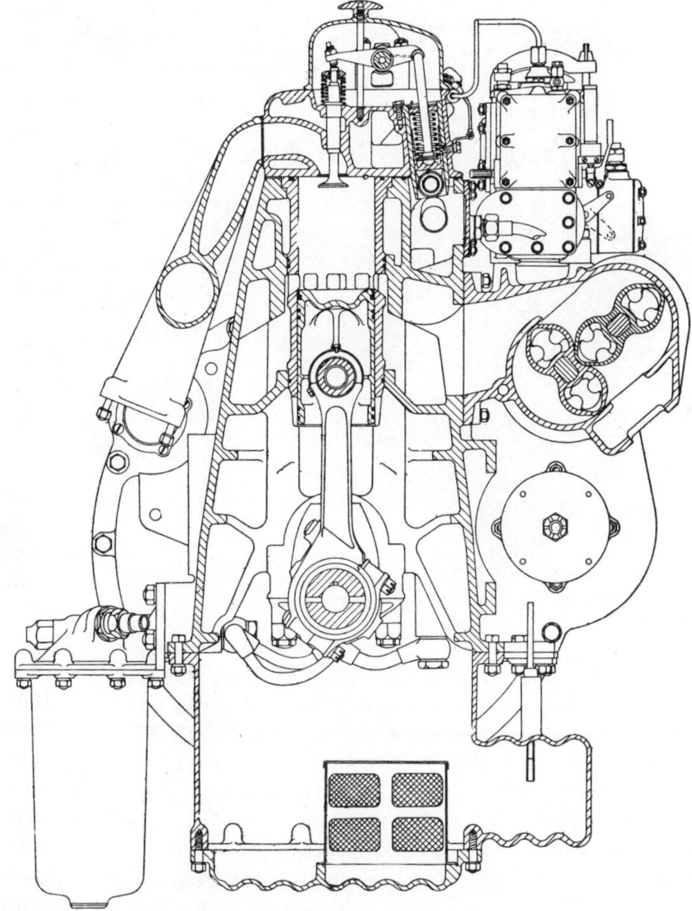

End section of the Foden two-stroke diesel engine

dead centre) the exhaust ports open suddenly with an explosive release of pressure to the exhaust manifold. The momentum of the column of exhaust gas causes a drop of pressure in the cylinder at the moment when the inlet ports open (about 44 deg. before o.d.c.) and this greatly assists the induction of the scavenge air as shown in the fourth diagram. Thus the Kadenacy effect is used to assist the work of the scavenge blower by the sudden release of pressure to the exhaust manifold, followed by carefully timed opening of the inlet ports. Both sets of ports are covered almost simultaneously at the start of the compression stroke at the moment when the reflex pressure wave has raised the pressure in the cylinder. The angular lead of the exhaust piston is obtained by positioning the axis of the crankshaft slightly higher than the centres of the pins at the bottom of the rocking levers.

In the case of opposed-piston engines which have two separate crankshafts coupled by a train of gears, the lead of the exhaust piston is obtained by

Off side of Foden engine, showing the exhaust manifolding

angular advance of the appropriate crankshaft. Examples of this type of engine are the 240 bhp Rolls-Royce K.60 and the Leyland 700 bhp L.60. Valve-in-head uniflow two-strokes also make use of this particular effect and here the sudden release of exhaust pressure can be assisted by the use of masked valves, the angular lead of the exhaust being obtained by appropriate timing of the valves. The Foden and the GMC two-stroke engines are well-known examples of this type.

Closely connected with combustion chamber design and combustion phenomena is the subject of forced induction or supercharging. As has already been explained, the maximum power obtainable by the combustion of fuel in the cylinder of a diesel engine is limited by the amount of air available. From

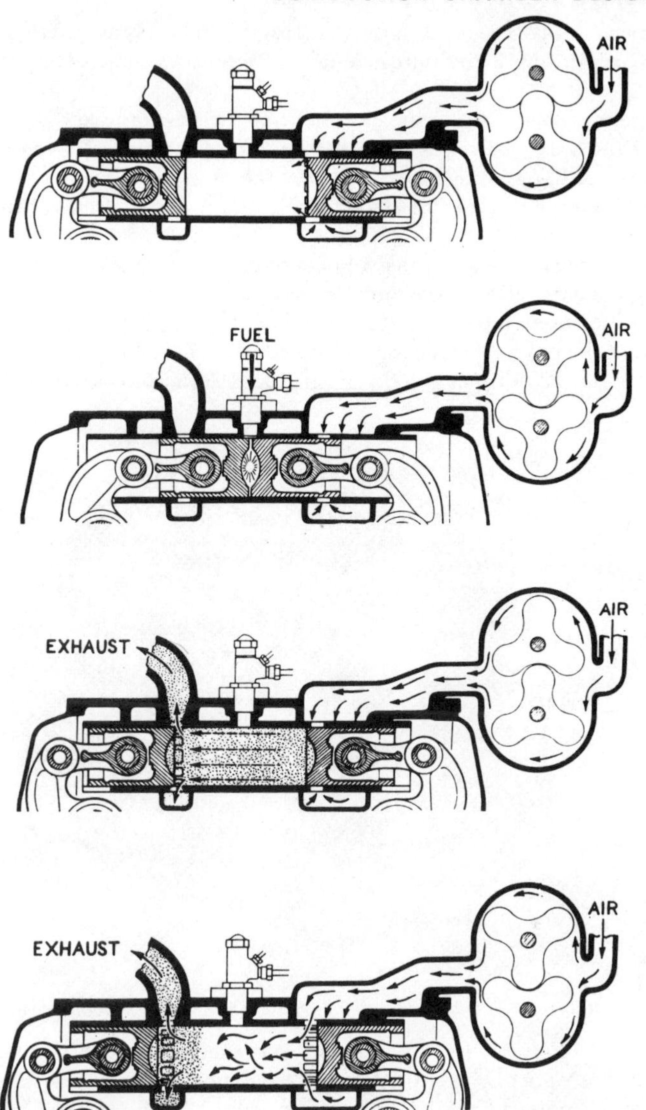

Schematic diagram showing cylinder layout and operational cycle of Rootes opposed-piston two-stroke engine

what has been stated earlier in this chapter, it is evident that the aim of the designer must always be the maximum possible utilisation of the available oxygen in the cylinder. But the cylinder is never completely filled to atmospheric pressure because of the inevitable drop of pressure through the inlet valves or ports, particularly at high engine speeds. The idea of assisting the natural aspiration of the engine by means of forced induction or supercharging has long been attractive to the diesel designer as a means of getting more efficient combustion and more power from a given size of engine.

The diesel cycle lends itself particularly well to supercharging as the blower has to compress air only as against petrol-air mixture in the petrol engine.

For many years it has been common practice to fit superchargers to large stationary rail traction and marine engines, of both two-stroke and four-stroke types.

In the early stages of the development of the automotive diesel, the difficulties in the way of producing a satisfactory supercharged engine seemed formidable, although the advantages of the system were obvious. But the addition of a supercharger to a four-stroke engine was often regarded as an undesirable complication and expense, although the blower had always been accepted as a necessary and integral part of the two-stroke engine.

During the early 1930s experiments were carried out with a Leyland vehicle engine with a supercharge pressure of 0·35 to 0·42 kgf/cm^2(5 to 6 lbf/in^2) and showed that the power output was increased from 55 to 75 bhp at 1 000 rev/min and from 95 to 125 bhp at 2 000 rev/min. In 1938 a number of AEC double-

Cummins (USA) four-stroke direct-injection engines are available with gear-driven Roots-type supercharger

decker diesel engined buses were put into service by the Halifax (Yorks) Passenger Transport Department and fitted experimentally with Centric vane type superchargers. The increase in power from 130 bhp to 175 bhp by supercharging at 0·42 kgf/cm^2 (6 lbf/in^2) enabled these buses to operate much more satisfactorily on the steep gradients in the town.

There are two main types of supercharger available to the designer, in one type the air is compressed by being positively displaced and in the other type the air is compressed by a high-speed centrifugal fan or impeller. The positive displacement blower may take various forms, such as the reciprocating pump and an example of this is the large-bore, short-stroke pump fitted to the opposed-piston two-stroke diesel engine developed in Switzerland by Sulzer. In this engine the reciprocating pump is driven from one of the rocking beams. The eccentric vane type is another form of positive displacement blower, in which a number of radial vanes are driven by a slotted rotor mounted eccentrically in a cylindrical chamber. Yet another form is the well known Roots blower in which the air is positively displaced by a pair of rotors each fitted with two or more intermeshing lobes, the action being similar to that of a gear pump. The rotors are positively driven by external gears which ensure that rubbing contact between the lobes is avoided. Fine clearances between the tips of the lobes and the casing and between the intermeshed rotors are essential to avoid excessive leakage losses. This form of blower is a familiar feature of many two-stroke engines; the Foden is fitted with a two-lobe blower, while the GMC two-stroke and the Chrysler opposed-piston two-stroke are both fitted with three-lobe blowers. A feature of the Roots blower is the distinctive noise caused by pressure pulsations, and on the GMC engine helically lobed rotors are used to reduce this effect, while some designers favour inclined ports in the blower casing to achieve the same effect. A further example of a positive displacement blower is the elegant design advocated by Sir Harry Ricardo, in which 14 double-acting pistons are arranged in a circle around an axially disposed shaft upon which a Z-crank and a wobble plate mechanism are centrally mounted. There are thus 28 impulses per revolution. Intake and delivery sequences are controlled by rotary valves. This device is manufactured in several sizes by Wellworthy Ltd.

Apart from minor changes of volumetric efficiency with change of speed, the quantity of air delivered by a positive displacement blower is proportional to its speed, and it is this characteristic which makes it attractive for fitting to vehicle engines because the volume delivered can be exactly matched to the volume required by the engine throughout the speed range, thus ensuring a flat torque curve. It should be remembered that these blowers will absorb 20–30 bhp for an engine in the automotive range, and the problem of finding a suitable drive has been solved in various ways. V-belt drives were used on some of the early conversions but gearing provides a more compact drive and is now generally favoured.

In 1964 the Perkins Group executed development work on their six-cylinder 100 mm ($3\frac{7}{8}$ in) bore × 127 mm (5 in) stroke DDE engine in which a direct injection toroidal combustion chamber is employed. A system for differential supercharging has been developed whereby the engine is enabled to produce a rising torque characteristic as the engine speed drops with increase in load. A screw-type positive displacement supercharger is driven through an epicyclic differential-gear train interposed between the engine and the output shaft of the two-speed gearbox which incorporates a torque converter. This arrangement senses the torque required by the road wheels and adjusts the amount of supercharge automatically.

Turning now to the other main type of blower, this takes the form of a high-speed centrifugal impeller which must be driven at speeds well over 50 000 rev/min in the sizes suitable for automotive engines, and we have the problem of obtaining a suitable drive from the engine. This appeared insoluble until, in the early 1900s, Dr. Buchi demonstrated that a satisfactory solution could be achieved by the direct coupling of the impeller to an exhaust-gas driven turbine, and as has already been mentioned this arrangement has been commonplace for many years on large engines. But it was not until after World War II that it was found possible to produce turbochargers in sizes suitable for automotive engines.

At first it had been found that the aerodynamic efficiency of small turbochargers was low and it was difficult to match their characteristics to the requirements of the high-speed diesel. Also the development of bearings suitable for very high speeds presented a problem and the cost of manufacturing many of the components was high. However, the development of the inward-flow radial gas turbine (in effect, a centrifugal compressor in reverse) and the adoption of flexibly mounted plain bearings (which have overcome the problem of oil-film instability or 'whirl' in high-speed bearings) have made possible the production of a suitable small unit. The design has undergone drastic simplification and the production of the turbine rotor by the lost-wax casting process, which obviates expensive machining operations, has led to the manufacture of turbochargers suitable for automotive diesels at competitive prices. These units are light enough to be supported on the engine exhaust manifold and they are lubricated from the engine oil supply, they are robust, and they do not require frequent maintenance.

The turbo-blower and the engine characteristics must be carefully matched to give a satisfactory overall performance, and it must be remembered that much more fuel is being burned in the same size of cylinder, causing a greatly increased flow of heat to the walls of the combustion chamber and a general

Holset model 3LD turbocharger. This sectioned view shows the inward-flow radial turbine on the left and the centrifugal blower on the right

increase in the working temperature of many parts of the engine. Thus it has been found necessary to develop engines specifically designed to meet these conditions; cooling systems and pistons have had to be redesigned and fuel pump drives have had to be strengthened to cope with the greater duty of the fuel injection equipment. Also it has sometimes been found necessary to change to heavy-duty lubricants of higher detergency.

There is one important feature which must not be overlooked. The maximum rating of a turbo-blown engine is likely to be determined by the limit of safe thermal and mechanical loading, rather than by the appearance of black smoke in the exhaust. It is therefore dangerous to tamper with maximum fuel stops as overloading can easily cause serious damage.

The design of small turbochargers has been assisted by the knowledge gained from the metallurgical and other problems encountered during the development of aero gas turbines, but the problems peculiar to their application to high-speed diesels have necessitated a lot of research. An interesting achievement is the production of a range of small turbo-blowers wherein relatively minor changes can match the characteristics of the turbocharger to the requirements of a particular engine. Changes to the profile of the compressor

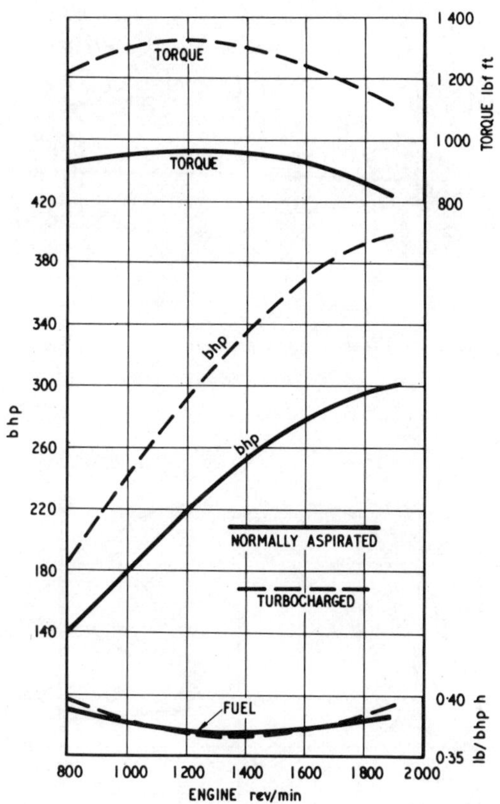

Comparative performance of AEC AV1100 (normally aspirated) and ATV1100 (turbocharged) 17.75 l. engines

wheel can give the desired air flow and the selection from a standard range of a suitable turbine nozzle ring will give the desired speed. The automotive diesel range is covered by a series of turbochargers having compressor wheel diameters between 76 mm (3 in) and 178 mm (7 in). It is startling to realise that these can produce a pressure ratio of 2·5 : 1 or more, with a compressor wheel of only 103 mm (4 in) diameter at speeds approaching 70 000 rev/min, with turbine inlet temperatures of 760°C (1 400°F), and can continue to operate with only minor attention throughout the major overhaul period of the engine. With these pressure ratios, the compression ratio of the engine is sometimes reduced to keep maximum cylinder pressures within reasonable limits, but this reduction is limited by cold-starting characteristics. An interesting development is the BICERI variable-compression piston described

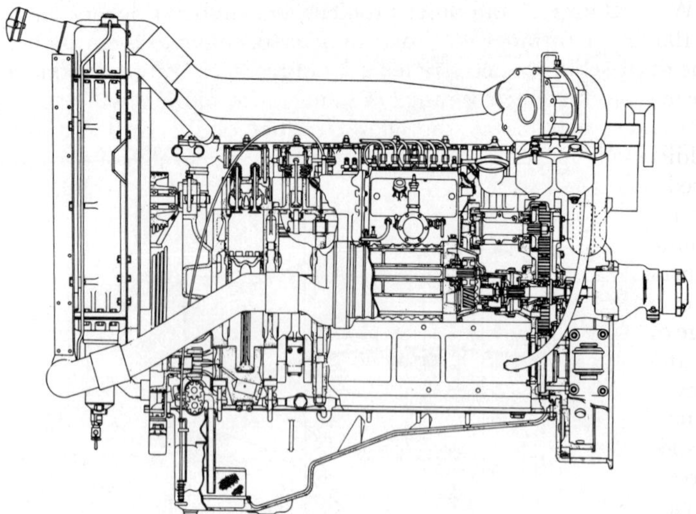

Sectional arrangement of Foden Mk VII turbocharged automotive two-stroke engine showing air-to-air intercooler fitted in front of radiator

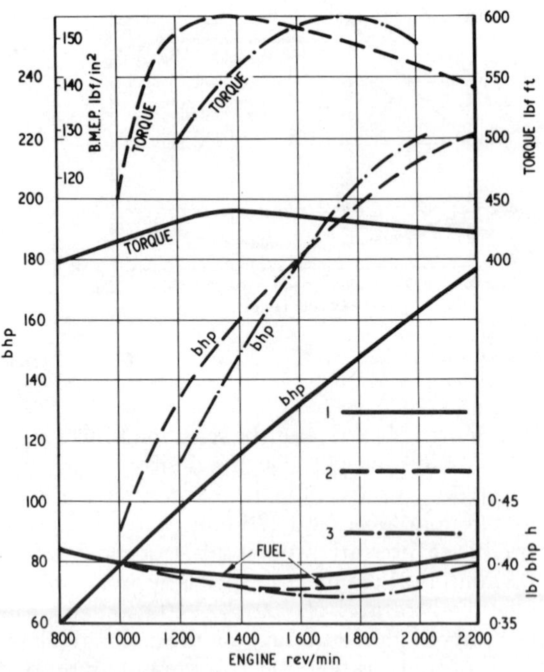

Comparative performance curves for Foden 6-cylinder two-stroke engines
(1) Mark VI blower scavenged
(2) Mark VII turbocharged automotive match.
(3) Mark VII turbocharged marine match

by Dr. W. P. Mansfield in papers to the Diesel Engineers and Users Association in 1959. In this design the compression ratio is increased by a supply of oil to the cavity between the two parts of the telescopic piston. The maximum pressure in the cylinder is controlled by a spring-loaded relief valve fitted to the oil cavity.

In addition to a 50% increase in the power from a turbocharged engine as compared with a naturally aspirated engine of similar weight and bulk, it is possible to achieve a 5% to 10% improvement in specific fuel consumption on account of the improved air/fuel ratio. There is thus a strong indication that a turbocharger will soon come to be regarded as a standard accessory on almost every automotive diesel engine. It is fair to say that the successful marriage of the gas turbine, with its ability to produce much power from small weight and bulk, to the diesel engine, with its unbeatably high thermal efficiency, has produced a power unit which combines many of the virtues of both prime movers.

It has long been the practice on large marine and stationary engines to fit an intercooler between the blower and the engine whereby the density of the charge can be increased still further by cooling. This system is particularly attractive where sea water can be used as the cooling medium, but for automotive engines air-to-air coolers are now coming into use.

An interesting feature of the application of turbocharging to automotive engines is the successful use of turbo-blowers in combination with Roots blowers on two-stroke engines. This not only leads to an increase in the power and the torque, together with a reduction in the specific fuel consumption, but by careful matching of the characteristics of turbocharger and engine the shape of the torque curve can be tailored to suit a particular duty. For example, an automotive engine can be induced to give a high torque at low speeds while the marine version can give a high torque at high speeds to match the propeller law.

7
Diesel fuels and lubricants

Like many new developments, the automotive diesel engine suffered from the effect of a number of exaggerated claims, not the least of these being that vehicles would be able to run on any kind of oil, even the waste oil from other machinery. That the engine *will* run on such fuel is a fact, but that such running is not advisable is another matter, for actually the high-speed diesel engine is as critical of its fuel as is the petrol engine.

Oil fuels for high-speed diesels are produced from the crude oil from which motor spirit is distilled. Those mainly in use are gas oils, distilled from the crude oil and closely related to the kerosene fractions. There is a considerable demand for gas oil for the enrichment of town gas, for the manufacture of petrol by cracking processes, and for domestic heating. Because the quantity of gas oil is necessarily limited, it seemed that the price would rise in view of the growing number of engines requiring it.

Diesel fuel is necessarily a refined oil, and at normal temperatures it is quite fluid and translucent—in appearance it closely resembles kerosene with a slight trace of lubricating oil. Such specifically prepared fuels have contributed in no small measure to the progress of the modern diesel, particularly in this country, where the wide use of the direct-injection engine has required the provision of the highest grade of oil fuel. In this country all petroleum fuels used for road vehicles are subject to taxation and the kind supplied for such use is named 'Derv' to distinguish it from the chemically similar 'gas oil' available for marine and industrial engines in which applications tax requirements differ.

In this connection a possible obstacle to diesel progress may one day present itself. Fuel for high-speed diesels is not residual oil in the sense that it is a somewhat unwanted by-product of petroleum distillation. It is a definite and particular 'cut' and the quantity obtainable per ton of crude is limited. Developments in petroleum technology have made it possible to extract more and more petrol by methods of breaking down the chemical structure of the heavier components and re-forming them; thus petrol yield still further reduces the amount of diesel oil. This is not to suggest that in time there may be no diesel oil, but rather to make it clear that its price must always be tied with the price of petrol and it may even become the more expensive fuel. Those who still think that the high-speed diesel has scope because it uses a 'cheap' fuel may well ponder the remark common in oil circles that 'it all comes out of the same barrel!'

For automotive diesel-engine fuel the most important quality is its combustion response to the compression temperature attained; oils of paraffinic base are best, while those obtained from aromatic crudes are the least satisfactory in this respect. The spontaneous-ignition temperature of oil derived from coal is in every case higher than that of the least responsive petroleum oils.

Very little information within the scope of the non-technical user is available even today on the matter of the ignition rating of diesel oils. The ordinary user of petrol may talk quite glibly of 'octane rating' at least knowing that a petrol of about 80-octane is usable in the average motor car even though it may be prone to pinking. He also knows that 90-octane can be regarded as an excellent anti-knock spirit while 100-octane is equivalent to a fifty: fifty mixture of petrol and benzole. Furthermore, it is common knowledge that exceedingly small additions of tetra-ethyl lead to comparatively low-grade petrol will produce these satisfactory anti-knock ratings.

Whereas it is desirable to reduce the self-ignition properties of petrol, the self-ignition quality of diesel fuel is a vital requirement, and is defined as the cetane number rating. Much research has been carried out in an attempt to develop suitable additives for improving the ignition quality of diesel fuel in the same way that tetra-ethyl lead can improve the octane rating of petrol. Unfortunately, the most suitable additives are detonants themselves and somewhat dangerous to handle in concentrated form. Also, they are volatile liquids and would lower the flash point of the fuel with an increase of the fire risk. It may be that research on this subject will lead to the production of premium grades of diesel fuels in the future but at the present time there does not appear to be any commercial advantage in the use of additives for the improvement of ignition quality.

This quality appears to be largely a matter of the selection of the most suitable crude oil and refining it correctly. Chemically stable fuels provide the most suitable all-round characteristics. In an attempt to increase the yield of diesel oil from crude oil, diesel fuel has been produced by a catalytic cracking or re-forming process, similar to the process used for increasing the yield of petrol. In practice the cracked fuel can be unstable and less suitable as a diesel fuel although its chemical and physical properties are very similar to those of a straight-run distillate.

In recent years legislation has been introduced in many countries limiting the amount of smoke allowed in the exhaust of vehicles; permitted smoke density levels have been specified and the type of instrument to be used has been stipulated. In some countries fuel additives are now marketed which have the effect of reducing exhaust smoke. They are claimed to be effective when added to the fuel in concentrations of about 1% but at present price levels this can add about 10% to the cost of the fuel.

Oils from coal, since they are generally of the unstable class, have not been a great success for diesel engines, although from time to time claims have been made, particularly with regard to oils derived from low-temperature distillation; these have been confirmed only by immediate results and not by long-term usage. Diesel oils have been produced from the distillation of Scottish shale; these are quite satisfactory, but the quantity available has never been other than negligible in relation to our total oil requirements.

Ignition quality is related to spontaneous ignition temperature (s.i.t.) and is indicated by the cetane number of the fuel which, for high-speed engines, must be about 50. Assessment of this figure is difficult; it requires the use of a specially equipped test engine and calls for skill on the part of the operator. There are several methods in use; in one method the engine is fitted with electronic equipment for measuring the lag, in terms of crankshaft degrees, between the first opening of the injector and the first measurable rise of cylinder pressure above the normal compression curve. In another method, the air intake to the engine is progressively restricted by operating a graduated throttle valve, and the suction depression is measured at the point when mis-firing is first observed. The measured ignition lag (or the measured suction depression) for the fuel of unknown ignition quality is then matched against readings obtained when using blends of test fuels of known cetane number.

It is notable that the British Standard specifications for diesel fuel published in 1937 did not include a figure for cetane number but a revised specification published in 1947 incorporated the rating, an acceptable standard test method having been established in the interim. The B.S. specification designates diesel fuel suitable for modern high-speed engines as Class A1 (see table). There is also a Class A2 fuel and this is a slightly lower grade suitable for

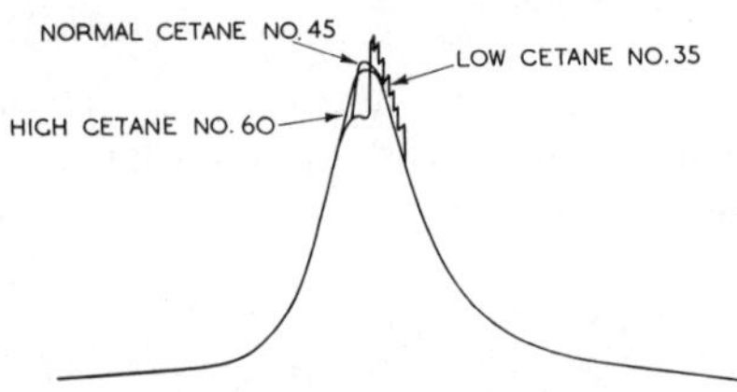

Effect of cetane number on combustion pressure. A low-rated fuel of poor ignitability produces a violent pressure rise (diesel knock) and unstable burning

industrial moderate speed engines as distinct from high-speed automotive engines. From time to time attempts have been made to run high-speed engines on Class A2 fuel, but almost invariably greatly increased wear of pistons, piston-rings, and cylinder liners has resulted and this has been attributed to the abrasive nature of the carbon formed by some of the slower burning constituents of the Class A2 fuel.

Apart from chemical suitability, physical cleanliness of diesel oil fuel for high-speed engines is of paramount importance, because the injection equipment is a precision mechanism built to 0·0025 mm (0·0001 in) working clearances. Usually there are both a coarse strainer and one or more fine filters in the line between tank and engine. Provision is also made for raising the fuel to a reasonably high temperature before it reaches the injection pump by placing the main filter (or the final filter of a series) on or near the cylinder block so that it is heated to water-jacket temperature or thereabouts; this ensures that the fuel is of the correct viscosity when it enters the injection pump cylinder. It is also important to avoid corrosion damage to fuel injection equipment, due to water in the fuel, and filters are now available which can prevent the passage of water droplets as well as of solid matter.

In the early days there was a tendency for wax to separate out of the fuel, causing it to gel in the fuel pipes and to choke the filters. However, more highly refined oils are now produced and this trouble is experienced only during periods of exceptionally severe cold in the British Isles. For more severe climates where very low temperatures are commonplace, special grades of fuel are available.

In the wartime search for alternatives to mineral oil fuels for high-speed diesel engines, the possibilities of vegetable oils were not overlooked. Tests carried out in this country and abroad showed satisfactory performance could be obtained with engines running on various vegetable oils, such as palm oil, cotton seed oil, ground nut oil and soya bean oil, although careful filtering and

BRITISH STANDARD 2869:1967—REQUIREMENTS FOR ENGINE FUELS

	Class A1	*Class A2*	*Class B1*	*Class B2**
Viscosity (centistokes, at 100°F)				
min.	1·6	1·6	—	—
max.	6·0	6·0	14	14
Cetane number, min.	50	45	35	—
Carbon (Conradson, % by weight), max.	—	—	0·2	1·5
Carbon (Conradson, on 10% residue, % by weight), max.	0·2	0·2	—	—
Distillation (recovery at 675°F, % by volume) min.	90	90	—	—
Flash point (closed), min.	130°F	130°F	150°F	150°F
Water (% by volume), max.	0·05	0·05	0·1	0·25
Sediment (% by weight), max.	0·01	0·01	0·02	0·05
Ash (% by weight), max.	0·01	0·01	0·01	0·02
Sulphur (% by weight), max.	0·5	1·0	1.5	1·8
Copper corrosion, max.	1	1	—	—
Cloud point, max				
summer	32°F	32°F	—	—
winter	20°F	20°F	—	—
Pour point, max.	—	—	30°F	35°F

*In Britain available for ocean bunkers only

preheating to reduce their viscosity were usually found necessary, and a loss of engine efficiency up to about 15% was experienced. Increased world demand for vegetable oils as food, however, preclude their wide use as fuel.

During World War II, most British military vehicles were fitted with petrol engines as the logistic problems associated with the use of more than one type of fuel in the field would have been unacceptable. However, the advantages of the diesel engine were most attractive to the military authorities and after the war intensive research was carried out in an attempt to develop a range of compression ignition engines which would operate on either diesel fuel or petrol. This has been achieved and a range of engines has been developed in which either fuel may be used. When operating on petrol some loss of power has to be accepted and it is advisable to use petrol having a comparatively low octane number on account of the poor ignition quality of high-octane petrol. It is interesting to note that these multi-fuel engines are of the two-stroke, opposed-piston, uniflow scavenge type and that conventional forms of combustion chambers are used. Some redesign of fuel injection equipment has been necessary to deal with the poor lubricating properties of petrol.

A problem that presented itself in the late 1940s in connection with diesel fuel was an unavoidable increase in sulphur content. In the British Standard quoted it will be noted that 1% is allowable, for Class A oils, but in point of fact the amount was usually well below 1%. Unfortunately, changing world conditions compelled petroleum refiners to draw increasingly upon crude oils from sources not previously used and the sulphur content rose in consequence. An exhaustive examination of the matter was presented in a paper by Broeze and Wilson of the Shell Research Organisation before the Automobile Division of the Institution of Mechanical Engineers in March, 1949. The implications were that the highest specified sulphur content might become the norm and that it would be reflected in an increased rate of cylinder bore, piston, and piston ring wear and that sticking rings and corrosion of injectors would entail more frequent servicing. Fortunately, the 'alarm and despondency' created at the time were not confirmed by subsequent events. Sulphur content did increase slightly but still remained below 1% in the grades of fuel supplied for automotive engines.

Some increases in the rate of deterioration of the lubricating oil occurred with increase of sulphur in the fuel, but the extensive use of chromium-plated piston rings and the development of lubricating oils with chemical additives did much to combat corrosive wear. Also, the introduction of better methods for the thermostatic control of coolant temperatures did much to reduce corrosion damage, by avoiding prolonged running with engine temperatures below the dew point of the sulphur compounds formed by the combustion of the sulphur in the fuel.

Lubrication always has been a matter very closely linked with fuel questions in the case of automotive diesels. Since fuel oil is non-volatile any unconsumed by combustion in the cylinder may either become carbon deposit or find its way past the piston rings into the sump. Dilution of the lubricating oil was thus a troublesome fault with certain of the early engines, although it was not generally a serious factor in regard to operation providing the oil was changed fairly frequently.

Far more serious, however, was the excessive amount of carbon deposit in the head and on the piston tops, also the contamination of the sump oil by sludge and a new phenomenon, 'lacquer'—a hard, polished brown deposit on the piston walls. Sticking of the top piston rings was also rather prevalent. These troubles were usually traced to excess fuel which was incompletely burned or to chemical decomposition of the lubricating oil at the high temperatures involved. Faulty injectors, if not changed at once, were (and are) a common cause of piston ring sticking due to serious accumulation of carbon deposit in the ring grooves.

Another cause of the trouble arises from the mistaken notion that more power is to be obtained by increasing the amount of injection; this is only true up to a point. The quantity of air taken into the cylinder is the determining factor and the amount of fuel that can be injected for optimum power is the amount that can be burned in that particular volume of air, no more and no less. Thus any trace of black smoke in the exhaust under full-power maximum rev/min conditions indicates that excess fuel is being injected, carbon deposit is forming and sump oil is probably being contaminated.

In British engines the demand for the highest thermal efficiency (lowest fuel consumption) has produced exceptionally good combustion conditions so that dilution, carbon deposit and sludging do not present abnormally serious difficulties in a properly adjusted unit. Fairly frequent oil changes and the use of high-quality oils are sufficient to maintain the engine in good condition and to keep lubrication costs within acceptable limits. Moreover, in the case of large fleets it was accepted practice to clean the drained-off oil by passing it through a centrifugal filter, the reclaimed oil being used for all intermediate topping-up or for mixture with an equal amount of new oil.

American engines presented the same problems but in a more acute form. The air cell combustion chambers favoured in the USA and the high-speed two-strokes, together with the efforts made to obtain speed and power comparable with petrol engines, led to the American operator tolerating lower thermal efficiency with a consequent drop in fuel economy of about 25% below the British practice. Sticking piston rings, contamination and sludging were therefore considerable troubles which the American diesel engineer handed over to the oil chemists in preference to improving the combustion characteristics of his engine and reducing rev/min with some loss of maximum power output.

Thus, during the war, when a great deal of American special military equipment had diesel engines, notably GMC and Caterpillar, the new detergent oils known as HD (heavy duty) came into great prominence. These oils contain a chemical additive or detergent which acts in the same way as soap does with dirt; each particle is enveloped by a slippery film so that it cannot

Piston from Caterpillar diesel after 375 hours' run on straight mineral oil and after 2000 hours on HD oil

stick to its neighbour to coagulate into a mass of sludge—instead each particle remains in suspension in the general fluid body of oil. In this way the accumulation of carbon behind the piston rings, on piston skirts, on the crankcase parts, and in filters is prevented; the oil soon turns black but it does not lose its lubricating properties.

Detergent oils have been found to be a complete cure for sticking piston rings in engines previously prone to the trouble and remarkable improvement in freedom from sludge has been observed. Of course, HD oils are more expensive and may have to be changed rather frequently.

Experiments with detergent oils in British engines soon indicated a decided advantage in internal cleanliness, but the final decision on their universal adoption was delayed somewhat to ascertain if the increased price could be offset by savings in maintenance costs arising from the extension of mileage or time intervals. Even in the case of engines not unduly susceptible to the troubles which detergent oils were developed to overcome, however, there was an obvious trend towards their general acceptance.

One of the first problems to appear with the introduction of HD oils was the trouble experienced by those operators who drained off their straight mineral oil and replaced it with HD oil. The detergent in the HD oil rapidly removed any accumulation of carbon and sludge from the internal surfaces of the engine, with disastrous results such as choked filters and blocked oil passages. It is essential to start using HD oil in a clean engine.

Filtration is a special problem in connection with additive treated HD oils. It has been suggested, although not conclusively proved, that certain fine filters fitted to the engine, notably the bypass type with chemically treated element, not only remove carbon particles but also remove the detergent additive itself. A. T. Wilford, Chief Chemist of London Transport, showed diagrams in a paper read to the Institution of Mechanical Engineers (1947) indicating that he had found a rather high rate of additive depletion. Similarly those municipal bus undertakings which utilise oil reclamation schemes on a large scale found that centrifugal filtration removed the special qualities of

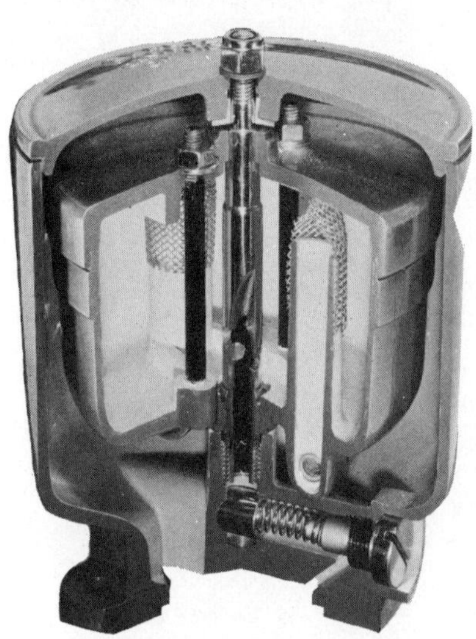

Sectional view of a Glacier centrifugal oil filter.

HD oil along with foreign matter and transformed it back into 'ordinary oil'. Still another problem was that of incompatibility, whereby the detergent oil of one blender would not mix satisfactorily with oil from another source. This trouble has now been largely overcome.

It should be remembered that the purpose of the detergent in HD oil is to hold in suspension all the carbon which would otherwise accumulate on internal surfaces. Large-capacity filters are necessary to contain this carbon and in this connection an interesting device is the centrifugal filter developed by the Glacier Metal Company in which the flow of oil is used to drive the bowl of the centrifuge. This item is now a standard fitting on some automotive diesels.

At one time, an increase in the consumption of lubricating oil dictated the period between major overhauls of the engine, but this was regarded as acceptable because the accumulation of carbon and sludge had to be removed periodically. The many improvements in the design of pistons, piston rings and particularly oil-control rings have greatly extended the period during which oil consumption can be expected to remain at an economically acceptable level and, thanks to detergent oils, the carbon is easily removed by changing the oil and the filter element. However, it is important to ensure that an adequate supply of oil always reaches the top ring groove so that the oil may irrigate the groove and allow the detergent to do its work by taking the inevitable carbon into suspension as it is formed. Excessive scraping can turn out to be false economy and can lead to a build-up of carbon in the top ring grooves. It is generally accepted that a satisfactory figure for the consumption of lubricating oil is in the region of ½% to 1% of the fuel burned.

It should be emphasised that the rate of additive depletion of the oil is related to the rate of carbon formation in the engine, and this is obviously related to the rating of the engine, the condition of the fuel injection equipment, and many other factors. It is most important to relate the oil change period to the operating conditions, as it is false economy to leave the oil in the engine after the additive has become depleted and is no longer able to do its work. A point sometimes overlooked with disastrous results is the importance of ensuring that stocks of HD oil are kept free from moisture due to condensation or leakage; some additives are water soluble and can be leached out of the oil.

It is now generally accepted that the greatly increased periods between major overhauls of the engine made possible by the use of HD oils lead to economies in operating costs which more than offset the extra cost of the oil, particularly when engines are highly rated or operate under arduous conditions. It is interesting to note that the makers of many of the highly rated engines now coming into service insist on the use of lubricants having much higher additive content than the HD oils formerly recommended for use with the more moderately rated engines of a few years ago.

Specifications have now been published for a range of HD oils of differing additive content. The laboratory assessment of the quality of these oils is an elaborate process. In addition to various physical and chemical tests, engine rating tests are carried out under carefully controlled conditions on special test engines. The quantity of carbon, lacquer, and sludge is assessed, any tendency to ring sticking is noted, the wear of piston, piston rings and grooves, liners and bearings is carefully measured, and any corrosion is also recorded.

As a sidelight on the general survey of the fuel situation as it affects the automotive diesel it is worth recording that during 1956–57 experiments were conducted on a fairly wide scale on the running of diesel vehicles on various mixtures of Class A diesel fuel and petrol. In Britain the reason for the investigation was the impression that owing to possible shortages of petroleum due to strained political relations in the Middle East, there was some risk that the refining companies might take the line of least resistance by processing the largest possible amount of petrol at the expense of gas oil; there certainly was a period when petrol was the more readily obtainable of the two and it was with the idea of obtaining maximum vehicle use that the scheme for 'stretching' the available Derv with petrol became attractive.

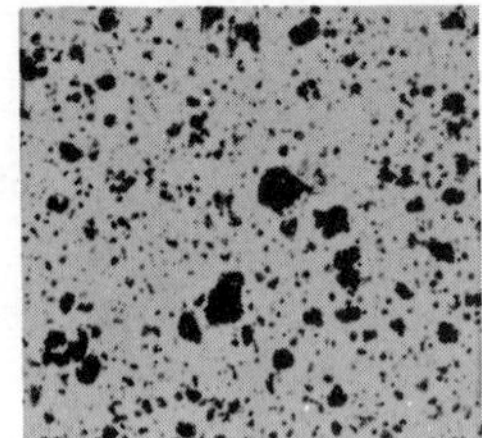
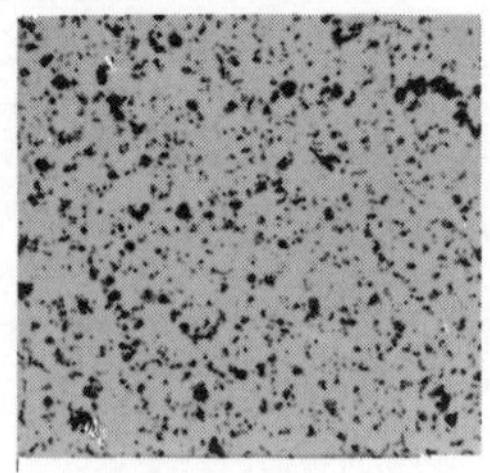

Photomicrographs of diesel fuel (× 100): (left) *contaminated by sludge and grit;* (centre) *after passing through felt block;* (right) *after passing through paper filter*

In Germany there were similar experiments, although the approach was from a slightly different angle. In that country it was thought that in the event of another war petrol production would again be expanded to the detriment of gas oil with the result that diesel vehicles would be at a disadvantage. It was found in both countries that normal diesel engines, apart from slight fuel pump readjustment, would run satisfactorily on average non-leaded commercial petrol and that even high-octane blends could be used.

During the 'Suez period', when fuel rationing was temporarily reintroduced in Britain, many operators, unable to obtain their full quota of Derv, did indeed stretch the quantity they had with petrol in proportions from 10% to 50%. Using 73-octane 'commercial' petrol, the Birmingham and Midland Omnibus Co. operated extensively on various Derv/petrol mixtures, the best results being obtained with rather less than one-third petrol. Starting was excellent but maximum power was down somewhat, while specific consumption was less favourable, as might be anticipated from the lower specific gravity of the mixture. The results indicated, however, that necessity rather than economics would dictate the use of a fuel of this type.

8
Standardising power measurement

After over six years of discussion a British standard for measuring the power output of automotive diesel engines (BS AU 141 : 1967) was at last settled. The next step is to negotiate an international settlement of what each manufacturer means exactly by a horsepower. At present, different methods of testing produce variations in claimed power output amounting to as much as 8%; an engine tested by one method might give 120 bhp, but by another method might give 130 bhp.

There has been a British Standard for several years specifying the conditions under which power output of diesel engines should be measured, but it has not dealt specifically with automotive applications, which is what the new standard does. The older standard for diesel engines was appropriate only for industrial applications.

To a considerable extent the automotive British Standard follows the rules laid down in a standard prepared for diesel-engine power outputs by the Society of Motor Manufacturers and Traders several years before. Previously, there had been variations in the inlet air temperature, the auxiliaries fitted to the engine when its power was measured, its exhaust smoke limit, the quality of fuel used, the tolerances allowed, the duration of running-in before a power measurement, and even in the interpretation of the terms 'net' and 'gross'.

The SMMT standard aimed at clearing up this muddle but in the end the standard received little recognition, mainly because it had no international standing and because it was entirely voluntary with no threat of disciplinary action if the standard was not followed. The SMMT standard was also a little unwieldy in that there were three ratings of engine power which were not easily discernible to the buying public and had rather long names that were not easily remembered.

Now, with the backing of the British Standards Institution, there is every reason for expecting that the new standard of diesel-power measurement will be observed nationally, and perhaps even internationally, because the motor trade of so many countries is now so interdependent in export markets. Moreover, it is well known that the settling of a British Standard is the first step towards the determination of a power/weight ratio for new vehicles and it is certain therefore that the British Government will be keeping a watchful eye on the behaviour of the diesel engine makers now that the new Standard has been published. Now most British manufacturers have re-tested their engines according to the conditions laid down in the Standard, and have published revised power outputs.

There are, it must be admitted, some practical difficulties in carrying out the BS requirements to the letter. Re-testing according to the selected conditions of test does not present much of a problem, but waiting for BS observation of a

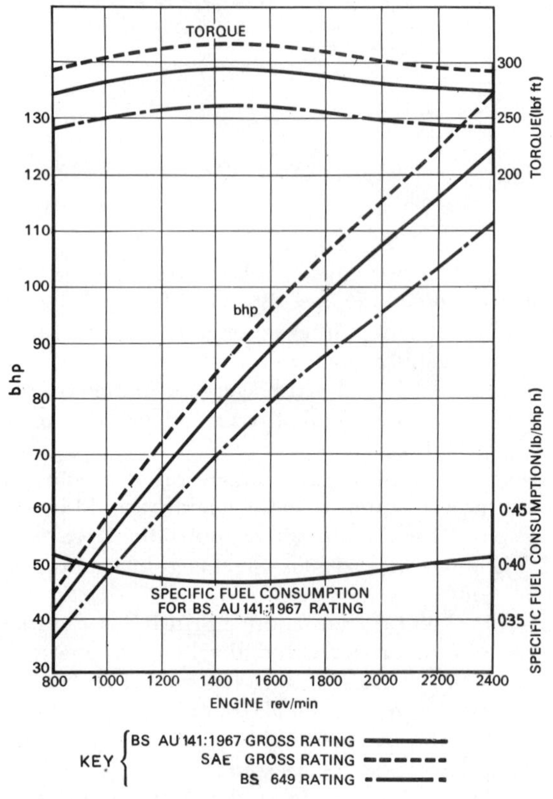

Comparative engine rating curves (Leyland 400)

100 h type test, now included in the automotive specification, is a costly and lengthy procedure which several engine makers are reluctant to undertake on every model of engine. It is taking time to fulfil this type-test section of the British Standard.

The aims of the British Standard are excellent. They provide the confidence of an independent body to give BSI type approval to an engine by taking a sample at any time from production, and checking that it gives not less than 95% of the published recommended output. This published output is determined by an initial test; if this is successful a type-test certificate is then issued by the British Standards Institution.

The British Standard for the initial test specifies the condition of the engine for conforming with the Standard, the test-bed conditions, the exhaust smoke density and details of the 100 h type test through which an engine must pass before it can receive a certificate.

It is on the 100 h durability test and the smoke-density limits that the British Standard is more severe than any other country's requirements.

There are two BS automotive ratings for diesel engines. One is termed the gross output, which is the power of a bare engine fitted with only the equipment and accessories it needs essential to its operation. The other rating gives the installed output, which is the power of the engine when all the auxiliaries are on it which would apply for whatever vehicle installation it is intended.

The gross output is the one with which engines are usually compared because this is the only reliable common denominator. The installed output varies from one type of vehicle to another and is useful mainly as a means of assessing vehicle performance. Installed power is the basis of British power/weight legislation for diesel-engined vehicles.

To be more specific, the gross output of an engine is measured without fan or radiator and without auxiliaries such as generator, pumps or compressors.

An automotive diesel test room at Shell's Thornton Research Centre. The final requirements of a wide range of engines are determined in this facility

For the installed output (net) rating the fan and radiator are included, and so is the vehicle exhaust system, or a system with equivalent back pressure. Auxiliaries are also included, although usually on no-load.

One British hp equals 1·014 metric hp (that means the French CV or German PS). For a straight translation of power, therefore, the continental engine makers give outputs which are slightly optimistic by British standards —for example, 200 metric hp is actually 197 British hp. This difference is swallowed, in practice, because continental methods of testing assume a net, not a gross, power. Later all powers will be quoted in watts, however.

The British Standard defines bhp as rev/min multiplied by the torque in lbf/ft, all divided by 5 250. Specific fuel consumption is measured in lb/bhp h —not pints. This affords direct correlation with the metric g/bhp h.

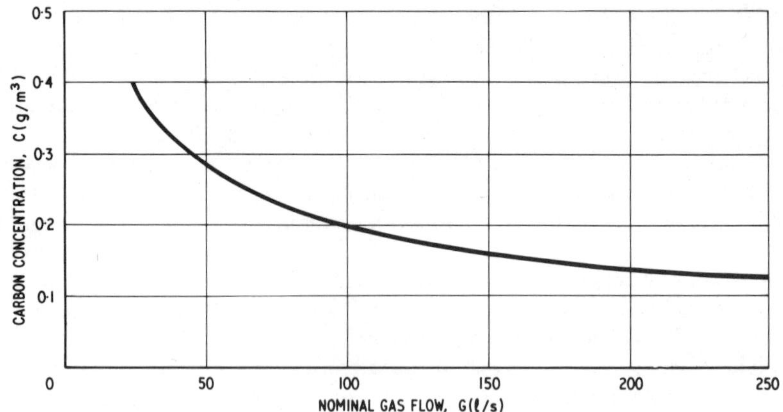

BS smoke/engine-size relationship chart

One of the modern engine-test cells at Shell's Darlington factory

As far as BS test-bed conditions are concerned, the barometric pressure is 29·92 inHg, the inlet air temperature is 20° C (68° F) and the calorific value of the fuel is 18 400 Btu/lb. No smoke suppressants are permitted in the fuel.

The British exhaust smoke standards need to be more generous as engine size decreases and so a formula has been devised to make it fair for all sizes of engine, except for supercharged engines where the gas flow is untypical. Exhaust smoke levels for supercharged engines are negotiated on the spot by British Standards Institution inspectors.

Exhaust smoke limits range from 65 Hartridge for very small engines to 30 for the very big. To the observer's eye, a 30 Hartridge smoke density is almost invisible. To take some typical examples, a 10–11 l. diesel is allowed only 34 Hartridge units at maximum speed, or 42 at maximum torque. A 6 l. engine is allowed 40 Hartridge units at maximum speed and 45 at maximum torque. A 2 l. engine is allowed 47 Hartridge units at maximum speed

A Hartridge BP smokemeter being made to measure the exhaust gas characteristics of a vehicle's diesel engine (Crown copyright)

and 60 at maximum torque. Often it will be the smoke limit at maximum speed which catches out some engines because of their shortage of breathing capacity at high speed.

Manufacturers conforming with the automotive British Standard have to publish curves showing the power output, torque, specific fuel consumption and smoke density. This is welcome, because several manufacturers at present shy away from publishing specific fuel consumption and smoke figures. Only the Hartridge BP and the Bosch-Dunedin smokemeters are recommended so far in the Standard.

The BS 100 h type test consists of 10 h running periods with at least 2 h standing time between them. Then, each 10 h running period consists of 50 min at three-quarters load at maximum speed, 45 min at full load and maximum torque, 5 min idling and 20 min full load and maximum speed. During this 100 h type-test, power, torque and fuel consumption must not fall by 5% compared with the initial readings. The same goes for the final readings. Two interruptions for adjustments are allowed during the whole test.

The aim of the British Standard is to settle a manufacturer's rating which will correspond with a 'satisfactory' service life and overhaul periods with 'suitable' maintenance. It looks as though the British Standards Institution has certainly taken great pains to achieve its ideal.

Attaining the next achievement—an international standard—is proving to be a long drawn out affair. However, an international standard on the actual test procedure has been virtually settled by the International Standards Organisation (ISO). The ISO proposals are broadly in line with the new British Standard except in one important respect, and that is the stipulation of exhaust smoke limits.

According to the ISO procedure an engine manufacturer will have to state smoke volume figures on his pass-out performance figures. This will be some guide for comparing engine performance technically, but does not tie every maker down to the same limits of smoke density, as does the new British Standard. For some time, therefore, the BS bhp of an engine may well be a slightly lower value, in practice, than the ISO bhp, even though the remaining test conditions are largely similar.

It is by the smoke limit stipulation that the British Standard penalises British engine makers. It is obviously easy to quote a high power output when no limit on smoke density has to be observed. But nearly all governments are

DIESEL ENGINE TEST CONDITIONS

	BS AU 141 (AU 141a 1970)	*DIN 70020*	*SAE J816a*
Engine condition	No fan, generator, pumps or compressor	Includes fan and unloaded generator and auxiliaries	No fan, generator, pumps or compressor
	No exhaust system	Includes exhaust system	No exhaust system
Smoke	Smoke limits 65–30 Hartridge	Smoke tolerable for road use	No smoke limit
Type test	100 h type test	No type test	No type test
Inlet temp.	68°F (20°C)	20°C (68°F)	85°F (30°C)
Atmos. pressure	29·92 inHg	29·95 inHg	29·38 inHg

keen to see legal smoke limits on vehicles; so even though smoke limits are not yet in the ISO standard, pressure will probably quite soon be brought by governments—at least the European ones—for limits to be incorporated. The means of doing this will probably be through the Economic Commission for Europe.

It is not the function of the ECE to prepare specifications for the purposes of regulations, but it can request the ISO to prepare a specification for quoting in subsequent legislation. Once the ECE makes a regulation, each member country is expected to incorporate the agreed law in its own national legislation. On diesel smoke limits all this is bound to take quite a long time however, and British engineers are resigned to its taking time to thrash out.

In the meantime, American engine builders test engines by the Society of Automotive Engineers (SAE) J816a test procedure; the Germans and several other continental countries test according to the standard DIN 70020 and the Italians have an IGM standard largely similar to the DIN (see table).

None of these takes account of exhaust smoke in power calculations, but engines to the DIN standard have to toe a reasonable line on smoke because

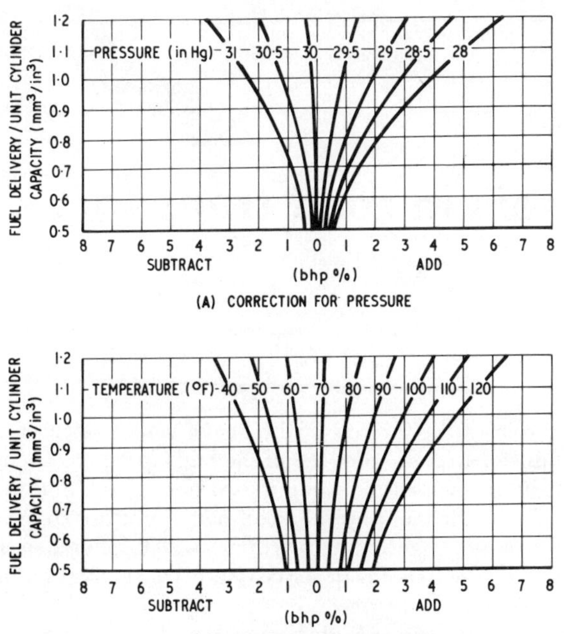

BS correction charts for adjusting quoted power according to barometric pressure and temperature for various sizes of engine (indicated by the fuel delivery per cylinder)

this standard refers to an engine in its installed condition, driving its fan and other auxiliaries. The American SAE figures are usually for a bare engine and so do not include the fan.

Test conditions go slightly against the SAE rating because the inlet air temperature is 85°F against 68°F (20°C) for BS and DIN, and the barometric pressure is 29·38 against 29·92 inHg, but SAE power ratings are still invariably higher than the BS and DIN ratings for any one engine—as a result of the more generous attitude on exhaust smoke.

Although DIN is a net rating and BS a gross rating, the fact that BS contains smoke limits usually detunes an engine by comparison. In the end the result is that the two power ratings are surprisingly similar for most engines, but there will never be a reliable basis for comparison, of course, until an ISO standard with smoke limits is agreed. That time is close, and there is good prospect that the same standard will be accepted in the USA, where great sensitivity is now being displayed on pollution. In fact the USA is setting a lead in tackling limitation of unpleasant chemical constituents in exhaust smoke—particularly carbon monoxide and oxides of nitrogen. Meeting the 1973 Californian standards has been no problem for diesel engines, although many engine makers are finding that the very stringent 1975 levels can be met comfortably only by turbocharging.

9
Diesels in road transport

It was mainly due to the interest of the heavy vehicle section of the British automobile industry that the progress of the high-speed diesel engine became so rapid. In 1930 heavy taxation was imposed upon goods and passenger road transport vehicles in Britain and this soon had operators and manufacturers alike looking around eagerly for anything which offered a reduction in operating costs. The search has never ceased because taxation has continued to increase as the years have gone on.

As fuel costs account for a high proportion of total operating costs (still about 20% in typical cases and over 30% in 1930) the improved fuel consumption which could be obtained immediately by switching from petrol to diesel engines has been of great consequence. In 1930 heavy lorries and buses typically had fuel consumptions in the 5–7 mile/gal (1·7–2·5 km/l.) range with petrol engines. The diesel engine offered between 9 and 12 mile/gal (3·2–4·3 km/l.) and there was the extra advantage in the early 1930s that fuel oil was not subject to the big tax which applied to petrol.

Of course, a tax equivalent to that on petrol was very soon also levied on fuel oil used for the propulsion of motor vehicles. Although this brought the actual consumption costs into perspective the economy represented just in used volume of fuel was still very real and was sufficient to justify continued interest in the diesel.

Unquestionably, it has been the diesel's fuel economy which has accounted for its becoming so dominant in commercial road transport that, in heavy vehicles at least, it has now displaced petrol engines completely in Europe. The attraction of low operating costs has always been great enough to offset the less pleasant characteristics of a diesel engine, particularly in respect of noise and weight.

Fresh situations regarding the rest of a vehicle's design had to be faced when diesel engines were first applied extensively to road transport. Taking a diesel and a petrol engine of equivalent power, a petrol engine will develop its maximum horsepower at a higher speed then a corresponding diesel. On the other hand, the diesel's maximum torque, or twisting force at the flywheel, is higher than that of a petrol engine of equivalent power and, moreover, this torque is maintained at a higher level over the whole speed range. This diesel characteristic, of low rev/min but high torque, made a fresh look at the vehicle's transmission necessary. The smaller speed range of the diesel meant that the number of ratios available in the transmission usually had to be increased. To absorb the high level of torque input the clutch and gearbox had to be made stronger. A diesel engine saves a little cost in only one direction; because it is thermally more efficient than a petrol engine it gives up less heat to its coolant and therefore less power is absorbed by the fan, and the radiator can be rather smaller.

Diesels could never have been a success in operation, of course, if they had not been reliable. They turned out to be not only more reliable, but also to need less maintenance and to have a longer life into the bargain. The diesel's fuel injection equipment was found to be less subject to derangement than the carburettor and electrical ignition apparatus of the petrol engine. The much sturdier construction, with generous bearing areas and heavier, long-skirt pistons, combined with the lower oil temperatures and comparatively cool exhaust, all helped to prolong the total life of a diesel engine far beyond what had become acceptable for petrol engines.

Noise and vibration, especially when idling, have always been problems with diesels and have never been solved successfully except at exorbitant cost (or sometimes by sacrifices in life and reliability). It is an inherent characteristic of the diesel cycle that there is a violent pressure rise during combustion. This occurs throughout the speed range but is more evident at low speeds. Besides this characteristic is 'diesel knock', the noise developed by the injection equipment, working at pressures around 2 000 lbf/in^2 (140 kgf/cm^2).

In a vehicle, the noise at high speeds is not so apparent as at low speeds because at high speeds other noises arising from the vehicle's motion push engine noise into the background. Nevertheless, the low-speed noise and vibration of a diesel engine have been factors in inhibiting its spread to ordinary car installations.

Fundamentally, however, diesels do not enjoy much popularity for light vehicles because of economics. The comparatively low annual mileages clocked up by the lighter vehicles make it hard for a diesel engine's fuel economy to yield sufficient cash saving to offset its lower specific power and greater weight compared with the petrol engines used at the lighter end of the

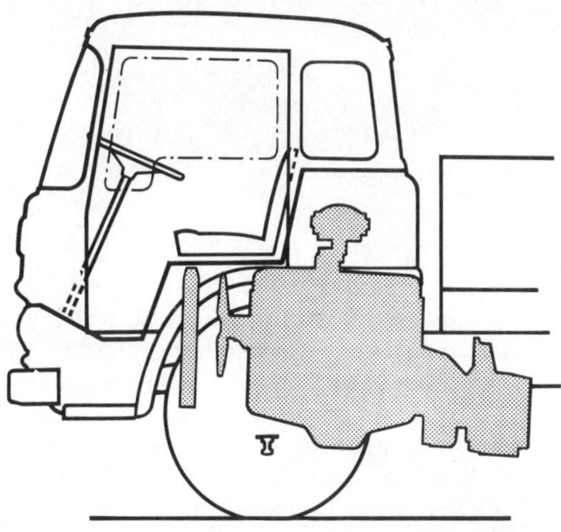

Bedford engine installation silhouette. In the modern truck installation a diesel's noise is reduced by mounting it well back in the cab

vehicle scale. It is therefore only in high-mileage light vehicles that it has been found worth while fitting diesel engines, unless diesel fuel happens to be very much cheaper than petrol, as it is in some countries.

Smoothing out the rougher idling of diesel engines has taxed the ingenuity of many a designer. Diesel engine characteristics are more violent and therefore a straight transfer of petrol-engine mounting techniques was not always very successful. After early attempts using spring-loaded mountings, swing links and rubber, the advent of bonded rubber components has finally successfully overcome the diesel mounting problem at comparatively low cost.

The more usual popular method of mounting an engine is now to suspend it at the front on bonded rubber sandwiches, inclined so that their perpendicular axes coincide on the centre of gravity of the engine—so that although a widespread mounting is provided it behaves geometrically as if the engine were supported at a single point. At the rear, bonded rubber sandwiches are sometimes used again, but more often bonded rubber conical mountings are used, which serve the purpose of locating the engine transversely and longitudinally as well as providing a means of cushioning.

Gardner flexible engine mounting

The rubber mix for these mountings has to be chosen carefully. It should be resistant to oil and the ravages of road spray and should have a natural frequency low enough to absorb most of the frequencies developed by the engine without build-up. In practice, say many designers, this means the rubber must be as soft as possible without deforming unduly under the weight of the engine.

Often, the choice of the correct mountings can be left to specialists, while the designer responsible for installation concentrates on the more familiar problems of spreading the load from the mountings into the vehicle's structure without encouraging fatigue cracking at the mounting points. When particularly soft mountings are used it sometimes becomes desirable to fit a small hydraulic telescopic damper between the engine and the supporting frame.

Designers of transmissions have had to follow diesel engine development closely. As the lower maximum speed of a diesel engine makes the vehicle performance more dependent on the way the torque can be harnessed, gearboxes with a wider spread of ratios have been needed. Gearboxes have had to be made together as well—and not just because of the high torque being put through them. There can be more subtle fatigue problems in gearboxes arising from the torsional oscillations in the engine and the general roughness in the power flow.

The weight of the engine flywheel has a marked influence on the transmission problems. There has been a tendency to lighten flywheels to make the engines more responsive to speed control and to reduce engine inertia so that it

One of the rubber sandwich-type resilient mountings at the front of a Perkins engine

slows down quickly when the accelerator is released and so makes rapid gear changes possible. However, if this flywheel lightening is pronounced the torsional oscillations in the output can become exaggerated and produce extra loading on the gears because of the constant acceleration and deceleration applied through the transmission at a high frequency. When gears are shot-peened their ability to resist this sort of treatment is enhanced, but a more fundamental cure is to arrange a torsion-bar drive in the clutch shaft.

The same problem of torsional oscillation can show itself in the clutch itself, where the main difficulty has been in obtaining a cushioned drive centre together with acceptable mechanical reliability.

A further clutch problem is arising now that engine powers are increasing beyond 200 hp for heavy goods vehicles. The stage has been reached where single-plate clutches are beginning to reach their limit of torque capacity and interest is being revived in twin-plate clutches which can take more torque yet have a small diameter and less inertia. A snag with twin-plate clutches is that it is sometimes difficult to ensure that both plates separate at the same time and that clutch engagement is always smooth.

Such clutch problems are tending to disappear as hydraulic couplings, in conjunction with epicyclic gearboxes, become more popular. Such transmissions have now been used in large numbers in buses throughout the world, the higher cost of this form of transmission being found to be more than offset by the greater reliability and ease of driving. It is now finding limited, but sustained, heavy loadings can affect reliability.

In the bus world interest is now extending beyond epicyclic transmissions towards automatic control. Electrical impulse methods have been used to control an epicyclic transmission automatically, but attention, in general, is being increasingly directed towards hydraulic torque converters which embody a hydraulic turbine giving an effective torque multiplication of the order of 2 : 1 or rather more, in conjunction with an epicyclic gearbox hydraulically controlled. There is more ready application of this type of transmission now that power/weight ratios of commercial vehicles have gone up

The bonded rubber mountings used for the back of a Henschel lorry; fore-and-aft rubber location pads are incorporated

and diesel engine speeds have risen. Such hydrodynamic transmissions have already been developed to a high pitch of efficiency in cars, and the experience is now being applied with good effect to commercial vehicle transmissions.

As faster diesel engines are entering the truck field more quickly than they are the bus field, there is a likelihood that in many cases the intermediate stage of development using semi-automatic epicyclic transmissions might be jumped as far as truck applications are concerned, and that there could well be a direct switch from syncromesh gearboxes to fully automatic transmissions.

There is a closer interest in automatic transmission for buses now and more operators are switching to one-man operation in which the driver collects the fares and issues tickets as well as driving. In such cases it is important that the

Installation of a ten-speed semi-automatic epicyclic self-changing gears compound gearbox in a Scammell three-axle tractor for 32 ton gross work

Rear-engine installations on buses, such as this Daimler, can give excellent access to the semi-automatic transmission as well as to the power unit

actual driving be made as easy as possible. Also, more significance is being attached to acceleration in bus performance characteristics these days, and again the automatic transmission has advantages here.

Some worsening of fuel consumption occurs as soon as a hydraulic coupling is incorporated in a transmission, but most designs of automatic transmission now incorporate a lock-up clutch which provides a slip-free drive from engine

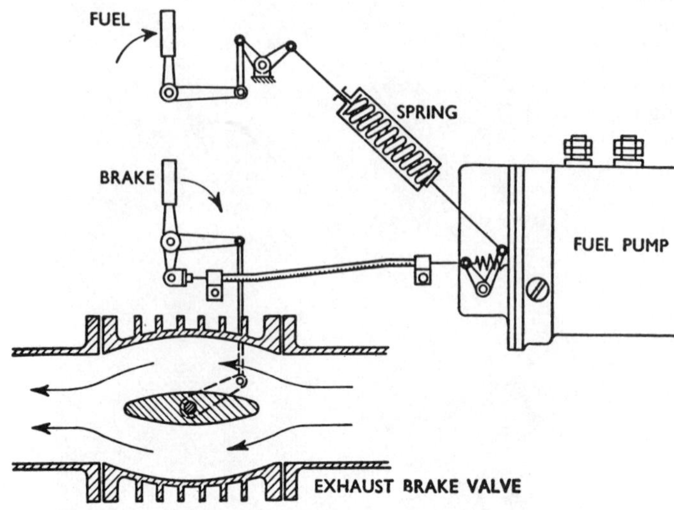

Diagram of an exhaust-brake layout

to gearbox at a predetermined engine speed or road speed. These refinements, together with increased efficiency from fluid couplings, have reduced the extra fuel consumption to a fairly minimal amount.

On all modern heavy and medium-weight road vehicles provision has to be made for power-assisted or power-operated braking, directly or indirectly derived from the engine. At one time braking was almost universally of the vacuum servo type, vacuum being produced by a sliding vane rotary exhauster driven in tandem with the engine's fuel injection pump. But now compressed-air braking is almost universal, and in this case the usual arrangement is to drive an air compressor from the engine. Sometimes the compressor is again driven in tandem with the fuel pump, but it is also often belt-driven on the engine, or shaft-driven remotely (especially on bus installations) or even driven from the layshaft of the gearbox. The compressors are usually single- or twin-cylinder units, rather akin to a small piston engine. Mostly they are air-cooled but lately there has been a growing trend towards water-cooled compressors, especially now that the demands on compressed air have gone up on commercial vehicles and bigger, harder-working compressors are needed. Air systems are bulky, of course, and there is revived interest in pressurised hydraulic braking systems for the future. Vehicles up to about 10 tons gross are the first to go into production with power hydraulics.

Additional braking power can often be provided by a four-stroke diesel engine if a cut-off valve is placed in the exhaust pipe, so that the exhaust outlet can be blocked and force the engine to work against pressure built up in the engine side of the exhaust system. Another method on the same principle cuts out the valve operation so the engine works as a pure compressor. This has been done in the past by the Swiss company Saurer and by the German company Krupp. The latest manufacturer to use the system is the Cummins Engine Co., in the USA and Britain.

With either system extra retardation is obtained from the engine on over-run, and this is especially useful on long descents. For this reason, gentle-retardation techniques such as exhaust braking have been adopted with more enthusiasm in parts of the Continent of Europe than in Britain and elsewhere where either the road conditions are substantially flat or else the hills tend to be comparatively short. The extra retardation obtained is rather similar to changing down another ratio in the gearbox—or, at the best, changing down two gears. Exhaust brakes have nevertheless been found beneficial to several operators in Britain where hilly districts have to be negotiated for a large part of the working time. Sometimes an unexpected side benefit has been automatic blowing off of dirt from paper-element air filters.

The latest auxiliary piece of equipment which is often needed on diesel engines for road transport applications is a hydraulic pump for the power assistance now being increasingly applied to steering. This is usually quite a

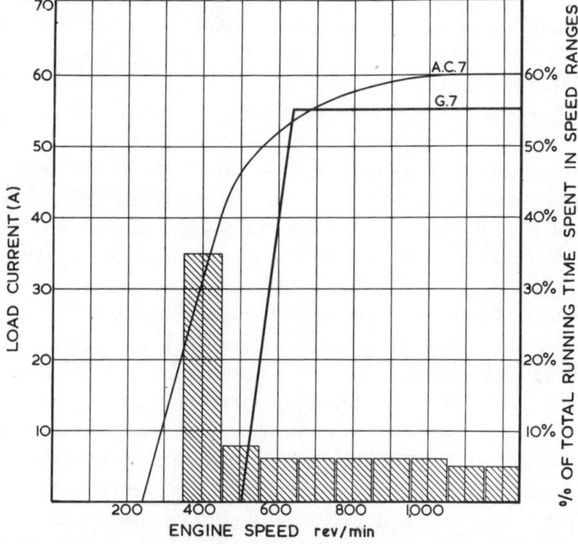

Comparative performance curves for a CAV alternator and dynamo

small pump driven fairly fast, the challenge being to generate enough hydraulic pressure at low engine speeds to be of benefit when manoeuvring at slow speeds. A special drive usually has to be accommodated for this pump.

For some work in the bulk transport sphere it is now often necessary to arrange a drive for a much bigger type of hydraulic pump, driven at or near crankshaft speed, in order to pump oil to a hydraulic motor driving some auxiliary piece of equipment such as a concrete mixer drum or a cargo discharge pump on a tanker. A most convenient place for mounting such a pump can be off the front of the crankshaft, but there can be difficulty with this type of installation on many vehicles because of insufficient space at the front of the engine; the power take-off on the gearbox is therefore often used.

Starting a diesel engine from cold requires some high sustained power from the electric starter motor and batteries. The regular band of starter power ranges from 4 hp for a 4 l. engine to 8 hp for the big engines with a capacity of more than 12 l. Progress in diesel engine efficiency tends to make it necessary to spin the engine faster for a successful cold start and this has brought fresh problems to the makers of starter motors. A starter motor also has to be robust enough to tolerate a steadily increasing current being put through it as the batteries discharge. The problem is being eased to some extent lately by the battery manufacturers altering the characteristics of the batteries so that more precedence is given to delivering a large amount of current at low temperatures rather than simply forming large reservoirs of electrical power.

This shift in priorities has been possible lately as a result of the big switch from direct-current generators to alternating-current generators for battery charging. The alternators charge more efficiently and charge effectively over a much wider speed range than dynamos. Battery design can therefore be concentrated on delivering starting current more efficiently. The need for plenty of power without excessively high currents brought about the general switch from 12 V to 24 V electrical systems on commercial vehicles of the heavier type. The switch to 24 V systems has brought problems of frailty in bulbs, although manufacturers in some countries have overcome this by having the lighting system on 12 V and the starting system on 24 V. Other methods of alleviating this problem of short filament-life are to use 28 V bulbs in a 24 V system or to use bulbs containing two 12 V filaments in series.

In the days of hand starting it was customary to incorporate a decompressor device on the engine so that the exhaust valves could be lifted temporarily while cranking speed could be built up at the starting handle. The technique was to release the decompressor when sufficient momentum was attained. The advent of reliable electric starting has made hand cranking obsolete, although the decompressor is still fitted to some engines in order to turn the engine more easily for adjustment and checking purposes.

On some installations a compressed air motor instead of an electric motor is used to start the engine, the advantage being the independence from electricity and greater mechanical reliability. However, air motors are more bulky than electrical ones of similar power, and considerable weight can be added to a vehicle as a result of the need to provide air storage capacity. Whether air starting becomes more popular really depends on the competitive standard of reliability which the makers of electric starters can provide.

Starting a diesel engine from cold is assisted when more fuel than is required for maximum power is injected while the engine is being cranked. To do this, in-line fuel injection pumps have an excess fuel device which allows the fuel pump to trip over to an amount beyond the normal maximum delivery. By British law the control for this excess fuel device has to be out of reach of the driver while he is the cab. Before this rule was made there were cases of drivers using the excess fuel to obtain more power up hills, but thereby generating dense exhaust smoke because such a lot of fuel was not being burnt. Engines fitted with distributor fuel pumps do not generally have an excess fuel device of the normal type, but they have a means of advancing the fuel injection to allow for the increased delay between injection and combustion at low air temperatures. Nevertheless, distributor pump designs are being developed which do indeed incorporate an excess fuel device.

All this starting difficulty is avoided by many bus companies and municipal undertakings by connecting up the cooling systems to a hot water or steam pipe on the parking ground. Another method is to use an electric immersion heater in the vehicles system, plugged into points on the parking ground. Heavy vehicles take up so much space that the cost of providing covered parking with central heating is usually prohibitive.

At extremely cold temperatures attachments to the inlet manifold are available in which a small quantity of ether is injected into the intake area. In the old days a favourite way of persuading a stubborn diesel engine to start in very cold temperatures was to hold a burning rag to the air intake. Nowadays the same principle is used in a more refined manner by having a red-hot electric element igniting a fuel spray in the inlet manifold. A very popular device of this type is the CAV Thermostart.

Heater plugs, looking rather like sparking plugs, but having a coil at the end instead of electrodes, are an accepted starting aid for indirect-injection engines where the combustion chamber is in the cylinder head. When current

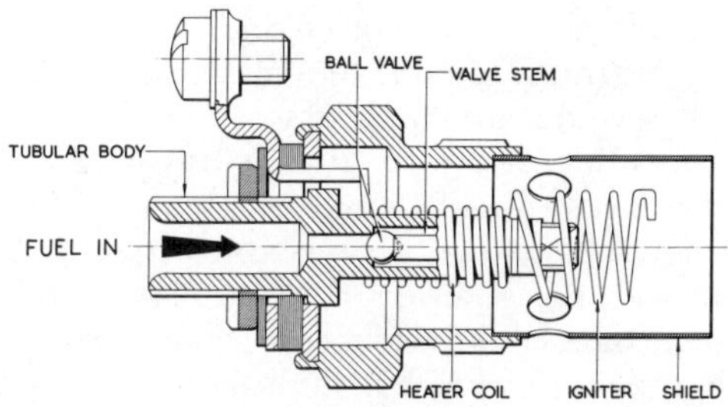

Section through a CAV Thermostart fuel burner for insertion in an inlet manifold to aid cold starting. Battery energy serves the igniter

flows to the heater plugs, which are wired in series, their elements at the tips glow red-hot and help to heat the air in the combustion chamber and to vaporise the fuel. Except in the smaller sizes of diesel engine, however, indirect injection is now rare.

Much of what has been reviewed so far in this chapter has concerned the effect diesel engines have had on vehicle design. But, equally, developments in commercial vehicles to suit their changing operational environment have in recent years had quite profound effects on diesel engine design.

One such effect has arisen from the higher standards of vehicle performance now generally required. Demand in overseas countries for higher power/weight ratios has been felt on European diesel engine producers for some time because of their substantial export trade. Extra impetus has been given to the trend to more powerful vehicles by the improved road conditions in Europe, and emphasis in national planning on better productivity in transport, meaning higher gross weights and faster speeds.

The big question is, 'How much extra power?' In West Germany and Switzerland the power/weight ratio set legally on heavy commercial vehicles has been 6 bhp/ton and this is the figure which (but using installed, not test-bed power) Britain is adopting. Already, though, the West Germans are demanding 8 bhp/ton. Power is the basic parameter of performance and is therefore a reasonable basis on which to legislate, even though it is rather rough and ready as far as heavy trucks are concerned because the type of transmission also affects performance in real terms.

But it can be argued that it is doubtful whether a legal minimum power/weight ratio is needed in practice. For one thing, even those vehicles which do not have enough power to attain the proposed minimum figure fall short of it by only a very small amount. Power/weight ratios hardly ever drop below 5 bhp/ton and a figure of 7 bhp/ton is quite frequent among the newer heavy machines now being produced. The average power/weight ratio is, in fact, about 6 bhp/ton, anyhow, and there is not much deviation either side of this figure.

Therefore, the ultimate effect on the usual speed at which heavy vehicles climb hills will actually be negligible, even where a power/weight ratio is legally enforced. Any increases in power at power/weight ratios of this sort do not produce more than a very small increase in hillclimbing speed. A 200 bhp vehicle might climb a 1-in-12 hill 25% faster than the 12 mile/h (19·7 km/h) of a 180 bhp machine, but anyone held up behind these vehicles going up a hill would hardly notice the difference. Getting an extra 20 bhp from a diesel engine typically means an increase of engine speed of about 400 rev/min, which is about 20% extra. If the maximum torque of the engine were not increased the vehicle would not climb any steeper hills but it would climb most hills 20% faster. But, of course, 20% extra speed does not amount to much when the basis is only about 15 mile/h (24 km/h). And it is when heavy vehicles are in the lower gears that they seem to attract public criticism.

The ultimate answer, unless distorted by legislation, is bound to have an economic root when commercial vehicle operation is considered. The future direction of big diesel development will, in the end, be determined by what the operators find is the best paying proposition. There are two distinct attitudes among British operators at the moment.

What influences these opinions is the high cost of fuel in Britain. Fuel consumption is still the largest element next to wages in operating costs, and many British operators are therefore appalled at the fuel consumptions of the extra-powerful lorries they have so far put into service. For example, they suddenly have to live with 7–8 mile/gal instead of the 9–10 mile/gal customary with lower-powered machines.

Therefore one section of operators (mainly hauliers looking for the lowest possible running costs) does not want much more power if fuel consumptions are going to be so poor. This section is thinking in terms of 6 bhp/ton.

The other section (mainly big concerns thinking of their public image) says it does want more powerful heavy lorries. It is thinking in terms of 8 bhp/ton.

From there on the pattern of future engines acceptable to the operating side of the industry begins to unfold.

In the first place, both sections of opinion are powerful and cannot be ignored in the future plans of any commercial vehicle manufacturer. More clues come to light when examining these two sections' other needs. Those who want the best fuel economy are also, on the whole, traditionalists, cautious and conservative in attitude. They would therefore prefer an in-line six-cylinder engine, which they understand and have lived with for so long, to any fresh layout full of apparently unknown quantities.

These men also place tremendous stress on reliability. With most of them being professional transport men, they are intolerant of breakdowns. They feel so strongly about this that they are prepared to sacrifice weight if this is a way of getting reliability and longevity. Certainly it should be appreciated that haulage is now much tougher on vehicles than it used to be. It is no longer rare for heavy vehicles to clock up 100 000 miles a year, leaving precious little spare time for maintenance.

With the section predominantly running vehicles privately on own-account, on the other hand, there is a preponderance of operators with apparently conflicting requirements. They say they want high power, but they also want vehicles as light as possible. And they think a lot about driver fatigue.

The implied challenge for engine designers is a stiff one: raise power outputs 70%, but do not increase the weight of the vehicle. The challenge is all the more important because it applies equally to many of the export markets of British manufacturers. There is, therefore, a double incentive to meet the challenge.

There is every indication that the answer to this problem of getting more power without increasing weight will lie in the V8 (or even V10 and V12). That this is a break with tradition does not matter much to the potential customers who are clamouring for the extra power. On the whole they tend to be more adventurous in their outlook.

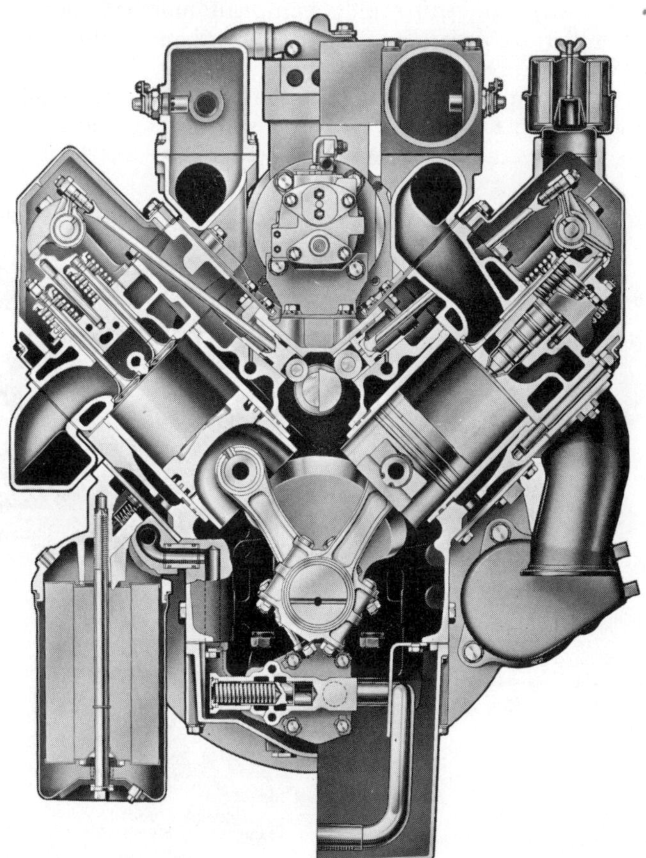

Cross-section of a short-stroke vee-form diesel with a peak speed of 3 300 rev/min—the 7·7 l. Cummins V8

The attractions of a V8 for power much above 200 bhp are that to get more power from a six-cylinder engine means making it bigger, or supercharging it, or raising the governed speed—assuming the engine's basic efficiency cannot be improved much more.

Making an engine bigger makes it heavier; supercharging (including turbocharging) does not add much weight, but raises internal stresses and temperatures. Either making it bigger or supercharging it raises the maximum torque, as well as the power. The extra torque requires a heavier clutch and a bigger, stronger gearbox, although not necessarily a heavier rear axle—provided the bottom-gear hillclimbing performance is not improved and provided the gross train weight has not been increased.

The only way of getting more power without increasing the engine weight and without stepping up the maximum torque and automatically requiring a heavier transmission is to raise the governed speed.

It is impossible to raise the governed speed of a big six-cylinder diesel very much without running into limitations of inertia loading and critical crankshaft vibration. Lighter pistons, shorter stroke, or smaller valves can reduce inertia loading, and having more cylinders to share a given capacity achieves these objectives. Crankshaft vibration becomes less critical when the crankshaft is made shorter and stiffer and, again, with more cylinders. Thus eight cylinders can increase governed speeds safely without spoiling the standards of long life and good reliability set by the slow in-line sixes.

The vee formation gives a stiffer crankshaft and valve gear and makes the engine short enough to tuck under the seats of the cab and improve driver

The Perkins 8·36 l. V8 diesel develops 170 bhp, but keeps the torque to 380 lbf ft

comfort at the same time. Modern tilt cabs provide easy access to a vee engine in such a position. The same transmissions can be used as are used already. Thus, although there will be exceptions, the future pattern of diesel engines for heavy commercial vehicles seems to favour vees for powers exceeding about 220 bhp, but still six-in-lines up to around that power.

Engine speeds are in any case going up. Speeds of about 3 000 rev/min were considered very high at one time, but experiments such as putting ungoverned diesel engines in cars proved that engines could actually sustain

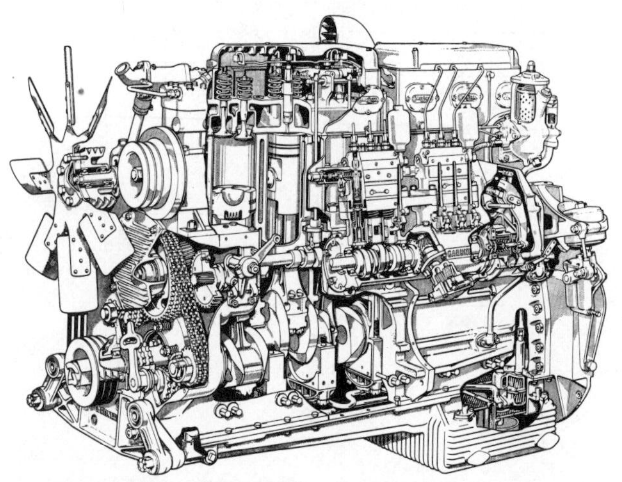

Still setting the standard in fuel economy and making British operators loath to depart from six-cylinder in-line type diesels—the 10·45 l. 180 bhp Gardner 6LXB

much higher speeds than usual for truck applications. Also, a great need arose to design diesels to withstand substantial over-revving when drivers descended hills. This became particularly prevalent when motorways were built. So, in any case, engines have had to have more reserve built into them to cope with high rotational speeds. The rate of wear does not go up in proportion to the increased speed. Oils keep improving, and use of oil coolers is spreading. So more engine makers are raising governed speeds, usually gaining at least a 10% increase in power.

Raising the governed speed is thus a neat way of getting more power out of the same size of engine. The attraction of this method of power boosting is marred only by the frequent tendency for engine efficiency to drop off disproportionately as speed goes up, and for fuel consumption to be heavy. This is why the governed speed of one 7·6 l. engine was clipped to 2 800 rev/min from the 3 000 rev/min for which it was designed. Rationalisation of parts, including valves, with a smaller engine meant sacrificing some breathing capacity at the top end of the speed range, causing the specific fuel consumption curve to rise rather too sharply for the last 200 rev/min.

At a premium, such problems can be overcome. For example, the Gardner 6LXB extracts from exactly the same 10·45 l. of the earlier 6LX 20% more power and 10% more torque. Despite about a 10% rise in full-load governed speed the fuel consumption curve remains remarkably flat. The specific fuel consumption at governed speed is, at only 0·336 lb/bhp h (152 g/bhp h), better

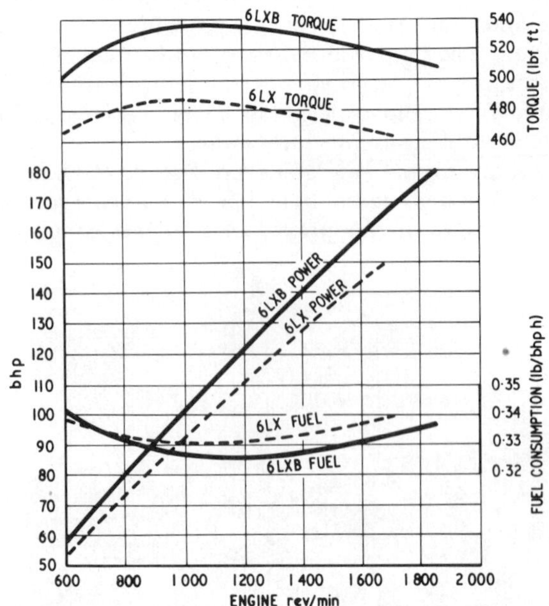

Comparative full-load performance curves of Gardner 6LX and 6LXB 10·45 l. engines

The 13 l. V8 designed by the AEC section of British Leyland

than almost any other comparable diesel gives at its best point. At its minimum level, around maximum-torque speed, the Gardner 6LXB's fuel consumption achieves 0·328 lb/bhp h (148 g/bhp h)—which is even better than the 6LX.

The Gardner 6LXB has two other big attractions. Its torque never drops below 500 lbf ft from 600 rev/min to maximum speed; and it weighs only 1 580 lb, which represents over 5 cwt advantage compared with the 200-bhp-plus engines from most competitive manufacturers. Even better specific figures are obtained from the 8LXB in-line eight-cylinder version of this Gardner engine, which develops 240 bhp at 1 850 rev/min.

The lightest heavy-vehicle diesel is still the Foden two-stroke. It has reached 180 bhp with the standard 4·8 l. six-cylinder models by raising the speed a mere 50 rev/min, which has produced the necessary extra 5 bhp over its previous power output. In addition to the standard Roots lobe blower to scavenge the cylinders of exhaust gases, by turbocharging Fodens can get a remarkable amount of power from 4·8 l. and 1 420 lb (644 kg)—225 bhp.

The Foden is a striking example of how high power/weight and power/volume ratios can be obtained by harnessing the two-stroke cycle. But there are two problems with two-stroke diesels which are sometimes sufficient to modify initial enthusiasm in them. One is the thermal stresses which concentrate around the combustion. The other is their high cost. There are other

In the new Bedford diesel engines thermal stresses in the cylinder head are eased by alternating the exhaust and inlet ports. Also, the water flows from one end of the head to the other, without any vertical cross-passages to the cylinder block

problems such as the rapid fall-away of torque at low speeds and the diminished retardation available on overrun, but these are minor compared with those concerned with cost and predictable reliability. Nevertheless, these problems can be solved, as witness the success of the General Motors two-stroke diesels.

Two-strokes become more competitive as powers go up. Their big attribute of compactness shows to enhanced effect when comparing engines above 200 bhp until, at 300 bhp, the two-stroke probably has a value advantage compared with a four-stroke diesel of this power. As vehicles become more powerful, therefore, the economics of two-stroke diesels become particularly attractive.

The thermal stress problems are harder to solve. It is only to be expected that if something like 50% more energy is developed in a cylinder, the temperatures are going to be higher. This makes it much harder for designers to keep piston expansion within bounds and to match the coolant flow to the hot spots without overcooling other sections of the engine. Design of castings is critical, because the bigger temperature discrepancy between hot and cool parts causes severe temperature gradients, promoting 'fighting' within the metal between parts which want to expand and parts which do not. It is not unusual for cracks to develop as a result of these thermal stresses, particularly in cylinder heads. There is a process known as 'tuftriding', which encases the casting in a skin, and puts the component in a state of compressive stress, which discourages cracking. Another approach to the problem is to use a reasonably high-tensile aluminium alloy which will conduct heat through its material before any damage can result.

With the Chrysler three-cylinder two-stroke diesel (previously the Tilling Stevens TS3 before Tilling Stevens was absorbed into the Chrysler-controlled Rootes Group) there is no cylinder head problem or even a valve problem, because two opposed pistons reciprocate in each cylinder bore. Combustion takes place between the heads of the two pistons when they come together at the tops of their strokes. The effort from each piston is transmitted to the crankshaft via an oscillating rocker arm at each side of the engine. This is an arrangement established some time before in the heavy industrial diesel field by the Swiss company Sulzer.

Another method of transmitting the power from the pistons of an opposed-piston engine is to provide a separate crankshaft for each set of pistons and then gear the two crankshafts together. This approach was first used successfully

The 700 bhp opposed-piston Leyland L60 two-stroke diesel is designed to run on a wide variety of light-distillate fuels. It has crankshafts top and bottom, geared together.

on a large scale by the German Junkers concern for its aero and industrial diesel engines, and has since been taken up by Coventry Climax, Leyland, Napier and Rolls-Royce for their opposed-piston two-stroke diesel designs, all of which are notable for producing a tremendous amount of power from surprisingly compact dimensions.

Whichever method of harnessing opposed pistons is used, however, the cost always comes out more than a four-stroke diesel of comparable size (though not necessarily more than for a four-stroke of comparable power). This is despite the elimination of valve gear and separate cylinder heads.

Cost is not a primary consideration when it comes to military applications, and it is here where the opposed-piston two-stroke diesel has found an important niche, at least in Britain. In its search for an engine which would run equally happily on diesel fuel, aviation kerosene or petrol, the British Fighting Vehicles Research and Development Establishment discovered that a Chrysler

Turbocharger installed on a current Leyland standard model—the fixed head '500' unit

opposed-piston two-stroke diesel would perform such multi-fuel duties with hardly any modification.

With the help of Chrysler engineers the Establishment instigated the design of a range of five opposed-piston multi-fuel diesels by Coventry Climax, Rolls-Royce and Leyland (in ascending order of size) up to 700 bhp. All are of the type having two crankshafts geared together. The Leyland and Rolls-Royce versions have since been offered on the civilian market, but they are very expensive at present. A Rolls-Royce version was applied by British Petroleum to an experimental rear-engined chassisless tanker for trial service on the Continent of Europe, but otherwise there has not yet been another civilian application of this type of engine.

Growing competition from two-stroke diesels is being met by turbocharging four-strokes, however. This is a compact means of boosting power by between 20 and 30%. By channelling the induction air through an air-to-air radiator (termed intercooling) even bigger power boosts are obtainable. Turbocharging has now reached the production stage on Cummins, DAF, FBW, Ford, Leyland, M.A.N., Perkins, Rolls-Royce, Saurer, Scania, Steyr and Volvo four-stroke diesels and has given encouraging results experimentally on Bedford and Gardner engines in goods vehicles.

Mechanically driven superchargers, though bulky and expensive, have often been considered a more reliable proposition, but the standard of reliability of modern turbochargers, such as Ai Research, CAV and Holset, is now so high that there are fewer qualms these days so far as the turbochargers themselves are concerned. Being sensitive to exhaust-gas volume and velocity they give good power boost response at low rev/min and large throttle openings —giving an uplift in torque when it is usually most needed. They are also self-compensating for atmospheric pressure changes, enabling power to be maintained in the rarefied air at high altitudes.

As with two-stroke diesels, turbocharging or supercharging involves more heat generated within engines and again this can bring thermal stress problems. Piston temperatures can increase particularly, and this is why Leyland, for example, arrange an oil spray from the top of the connecting rod to the underside of the piston in their turbocharged versions of the 6·5 l. 401, 11·1 l. 680 and fixed-head 8·2 l. 500 engines. This last engine is unusual in having

The fixed-head, overhead camshaft Leyland 500

the head and block cast in one piece; the aim has been to eliminate gasket problems at high temperatures and pressures and to gain more freedom in port design. It is also able to rev faster by having an overhead camshaft.

Thermal-stress problems are always exaggerated in automotive applications because the work-demand, and consequently the temperature, is not constant. With industrial, marine and railway applications temperatures are more even. Turbocharging has therefore been successful in such circumstances where it has been insufficiently attractive for automotive duties. Work in this field has been done by AEC, Cummins, Leyland, M.A.N., Perkins and Rolls-Royce.

It seems certain that more widespread use of oil coolers will follow from the more common use of turbocharging. Oil coolers are already often being found advisable on naturally aspirated engines now that long periods of high-speed running are regularly met on modern road systems.

10
Railway service

Diesel traction is now commonplace on railways throughout the world. Many economic factors which accounted for the rise in popularity of diesel engines in road transport have applied also to the railways. It is not just the big diesel locomotives of up to (now) around 4 000 bhp which account for diesel traction as far as the railways are concerned, however. The automotive diesel engine has also found an important niche in railway service. These comparatively small (by railway standards) engines have been a means of introducing the railcar concept on a large enough scale to make many urban and rural services an economic proposition again, on stretches of track which could not carry heavy locomotives and remain viable.

The basic idea of a railcar is virtually to put a bus on railway lines. Consequently, within the standard railway limits for withstanding shunting and coupling, railcars are built on lightweight principles as far as possible and are powered by engine-transmission assemblies which follow closely the installation in an underfloor engined single-decker bus. By using broadly the same type of transmission, good fuel economy on stop–start services can be obtained.

The analogy with underfloor engined buses made existing bus engines of the horizontal type natural choices for many railcars, whether they worked solo or coupled to one or more other cars to make multiple diesel sets for local

Automotive units applied to rail traction

services. However, because of the heaviness of railcars compared with buses, and the need to spread the traction over more than one pair of wheels in order to give good acceleration without wheel slip, it became usual practice to equip railcars with two power packs having synchronised controls.

Diesel engine makers who had experience of underfloor bus installations were off to a flying start when it came to applying the same engines to railcars, although many hard lessons had to be learned in the 1930s. These engines were naturally attractive to the railways because they were already proven units on stop–start service, were in quantity production at a low cost, were compact and easy to service, and often had world-wide distribution organisations for spare parts. By adopting automotive diesels in conjunction with hydraulic torque converters or fluid flywheels with epicyclic gearboxes, a railcar could be built which had the advantages of a road vehicle: it consumed no fuel when standing by, was a clean subject in the running shed, was ready for service at any required moment, and was quickly refuelled, unobtrusively and at infrequent intervals. Easy exchange of power units was an asset in the running sheds.

Applications of automotive diesels to railway service have quite a long history. The Gardner LW series of engines were applied to rail service as early

Modified AEC–Ricardo bus engine and transmission adapted for a G.W.R. railcar in 1934

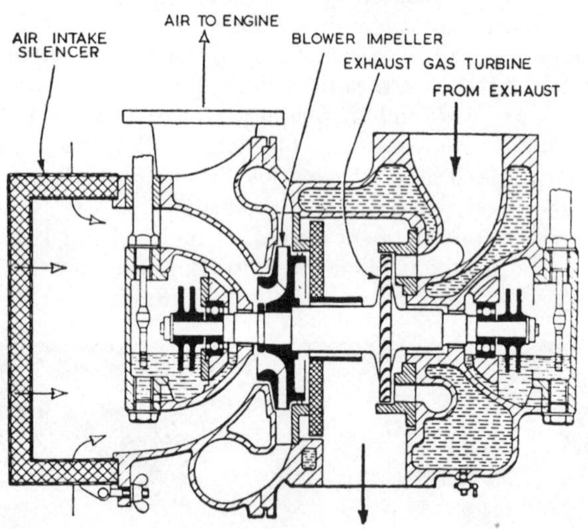

Sectional diagram of the Brown-Boveri exhaust turbo blower

as the 1930s in Argentina, Australia, Belgium and Ireland. In 1934, the Great Western Railway in Britain put a number of single-engined railcars into service powered by indirect-injection AEC six-cylinder engines developing 130 bhp at 2 000 rev/min.

Another unit supplied at this time by the automotive diesel industry for rail duties was a special Leyland 10 l. engine rated at 130 bhp. This was discontinued after a short while, but the smaller 8·6 l. road vehicle unit of 95 bhp at 1 900 rev/min had some rail applications, and in 1938 Leyland considered the application of the Buchi turbocharger system specifically for rail engines. Owing to the war the project was not seriously developed.

Development of trains powered by automotive diesels first got under way on a large scale in Northern Ireland. The experience of the Ulster Transport Authority (U.T.A.) in using AEC and Leyland horizontal bus engines formed a basis of subsequent application of such units to British Rail.

The AEC-engined cars of the U.T.A. had two 9·6-litre 125 bhp horizontal engines. These were essentially the same as those used for buses and they were mounted under the frame in the centre, but facing in opposite directions to drive the inner axle of each bogie. Transmission on these early units was by fluid flywheel and preselective epicyclic gearbox. Two powered cars (meaning a total of four engines giving 500 bhp in all) and an intermediate non-powered carriage formed a permanent three-car train to accommodate 250 passengers. The maximum speed was 75 mile/h. A later AEC development was to fit 11·3 l.

A 200 bhp AEC horizontal diesel of the type used to power many modern light rail vehicles of solo and coupled types

engines, which were big-bore versions of the 9·6 l. units, and developed 150 bhp. This power output was stepped up later in some cases by the application of the Brown-Boveri system of exhaust turbocharging. In this form the engine delivered 200 bhp at the same speed of 1 800 rev/min.

In the case of the Leyland-engined trains in Northern Ireland four twin-engined power-cars and two intermediate coaches form a six-car set of 1 000 bhp. Variations can be made from two-car to eight-car train sets by suitable arrangement of powered and non-powered cars.

For British Rail a similar arrangement was used in that two underfloor engines were mounted in the centre of the frame of the powered car. Whereas the U.T.A. Leyland-powered railcars had torque converters with a lock-up arrangement, the British Rail specification included electro-pneumatic control of epicyclic gearboxes.

Six-coach train of the Ulster Transport Authority. Four of the cars are each powered by a pair of Leyland 9·8 l. underfloor bus engines providing a total of 1 000 bhp. Maximum speed is over 70 mile/h

Installation of a B.U.T. power pack beneath a British Rail railcar

A British Rail multiple car set

Even before AEC Ltd. was absorbed into the Leyland group, the two companies set up a joint organisation called British United Traction Ltd. to coordinate the development of power units for railway use. One result of this marriage was the adaptation of a 15 l. Leyland six-cylinder engine called the O.900 which developed 230 bhp at 1 900 rev/min when normally aspirated and 275 bhp when turbocharged. In its original form this engine had wet cylinder liners, but later the arrangement was changed to dry liners. Another development occurred when Rolls-Royce Ltd. came on the railway scene with horizontal versions of the six-cylinder diesel engines it had originally designed for automotive and earth-moving equipment use. These have subsequently become popular with British Rail. There are also flat Paxman turbocharged sixes of 450 bhp at 1 500 rev/min.

Much railcar work has been done by Fiat in Italy, but in this case vertical engines are usually used. The Swiss diesel engine maker, Saurer, has also supplied light rail vehicle engines in various sizes, several with Buchi turbocharging.

In Europe supercharging has been more extensively applied to lightweight rail traction than in Britain. The exhaust-driven turbocharger, more suitable in its early days for constant-speed and fairly constant-load conditions than for the varying demands of road-transport applications, was widely adopted for railcars, particularly by Saurer in Switzerland and M.A.N. in Germany. Since those days turbochargers have been developed intensively in Britain for automotive use by CAV Ltd. and the Holset Engineering Co. Ltd., in particular, and these are both more responsive and more reliable than earlier designs of turbocharger. In the meantime Saurer has switched to a screw-lobe Lysholm mechanical supercharger and M.A.N. has also favoured mechanical supercharging until recently.

Diesel engines still provide the most economical form of power for rail traction, notwithstanding the glamour associated with gas-turbine trains and the public appeal of the highly capital-intensive electrified railway. Indeed, diesels could have shown their worth even more markedly if more use had been made of the best and longest proved automotive engines.

11
Tractor and industrial engines

Industrial application of high-speed diesel engines has now become big business and at both ends of the horsepower scale has extended to ranges of engines outside the normal span of automotive engines. However, there is no doubt that the development of automotive units had a big influence upon the application of diesel power to tractor, portable and industrial equipment. The economics of such engines are no less applicable to industrial than to automotive applications, and another big influence has been the reduced fire risk when diesel fuel is used as compared with petrol.

This Sheepbridge rock-breaker is powered by a Cummins NT400 industrial diesel

Special diesel engines built on automotive lines are now used, particularly in the agricultural field, on a big scale—usually in small four-cylinder form. At the other end of the scale, big units, still built on broadly automotive lines, are used in the heavy earth-moving equipment which is now commonplace in large civil engineering projects. But many pure automotive diesels still have wide use in de-rated industrial forms. They are used in the smaller earth-moving plant, in mobile cranes, for electricity generation, for portable or standby pumps, and for purposes in conditions where mains electricity supply is not readily available. For transportable plants their low weight is attractive.

Before 1939 the automotive type of engine had not extended much beyond its own immediate sphere, but after 1945 the increasing production capacity of the leading makers caused them to search for fresh applications for their units. Few of the concerns building heavy engines of relatively low specific output had the technical resources to cater for these new markets although they have since been influenced by the trend to lighter construction and higher speeds, as is evidenced by the appearance of industrial units of simplified form from the traditional makers of stationary diesels.

Obviously, in applications where compactness, availability of standardised servicing methods and ample stocks of spare parts are of consequence, the automotive diesel is particularly attractive. Furthermore, it provides a most useful form of portable power plant which can be handled as a compact and

A self-contained electricity generating set powered by a Perkins 6.354 industrial diesel

complete unit. This aspect has been rapidly developed in the form of self-contained power packs or the direct-coupled generator outfits for emergency or standby electricity supply.

In the earlier days engines of from 40 to 100 bhp were immediately available by de-rating existing automotive engines of the 70 to 125 bhp classes. This was done by limiting the maximum fuel delivery of the injection pump and by governing down from the normal 1 800–2 000 rev/min range to 1 500 rev/min or so. Since the early days, the power range provided by the basic automotive diesels has gone up as more powerful engines have come on the scene. It is now common for 200 bhp to be available from industrial engines based on automotive designs, and the success of turbocharging in industrial applications has enabled this power range to be extended beyond 300 bhp in some instances.

In many cases the automotive engine is being applied to industrial use without structural modification; the unit simply has its auxiliaries rearranged and

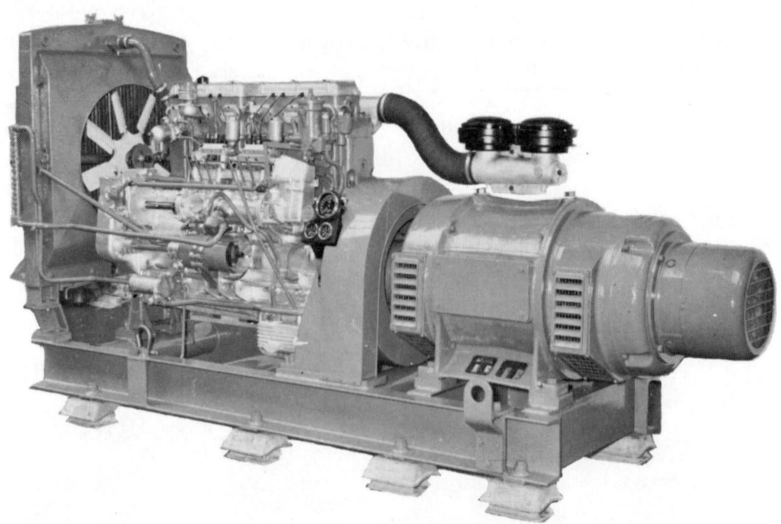

A Gardner 6LX engine coupled to a Brush 48 kW alternator running at 1 000 rev/min. The set is complete with a Gardner radiator and air-cooled oil cooler. The whole unit is supported on Metalastik Cushyfoot bonded-rubber mountings

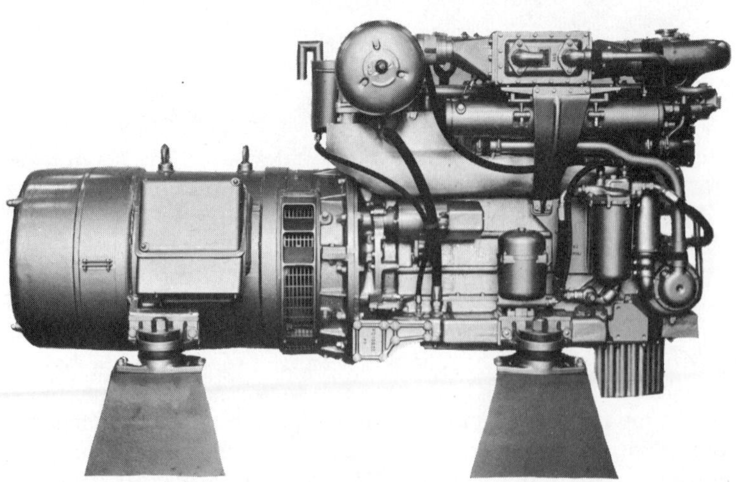

The turbocharged Foden two-stroke engine makes a compact generating set. Air from the turbocharger is cooled by an air-to-air heat exchanger before it passes to the engine's cylinders

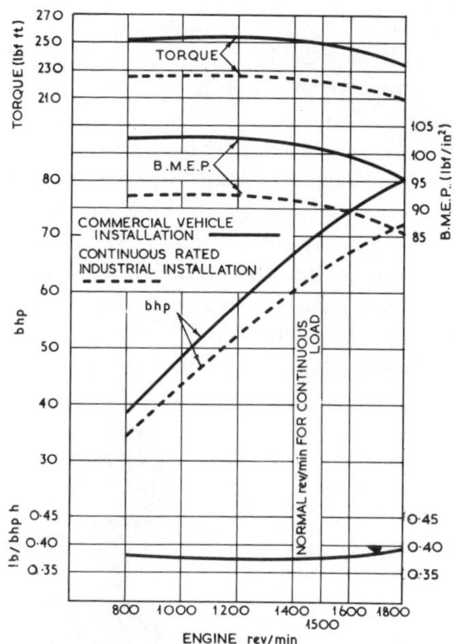

Performance curves of a 6 l. four-cylinder direct-injection engine showing difference between automotive and industrial ratings

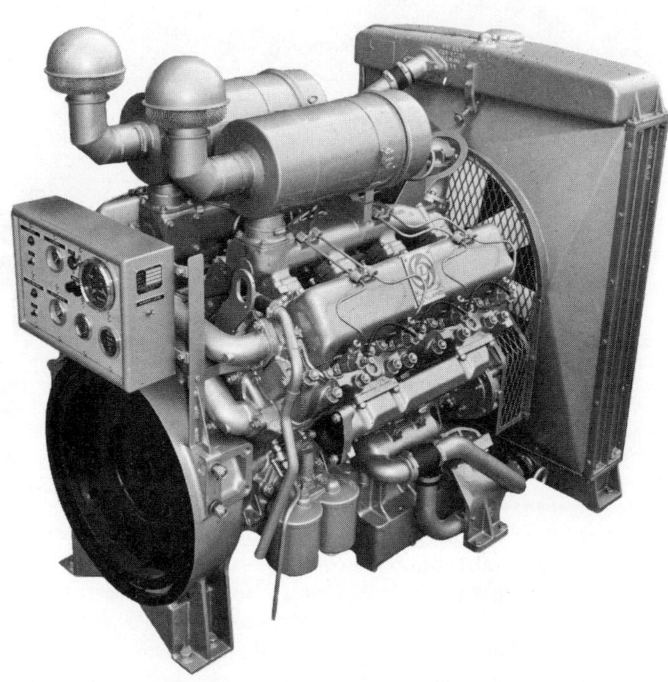

The AEC-designed British Leyland V800 12 l. V8 engine as a 230 bhp industrial pack

the whole engine is mounted on a bed-plate that has brackets which mate directly with the normal bearer facings on the crankcase. Often, the industrial engine business has become great enough to justify offering complete power packs enclosed in special cabinets.

Sometimes heavier flywheels are fitted to the industrial engines. For some purposes heavier flywheels are essential, in fact, particularly for driving alternators where cyclic variations exceeding the limits specified by the British Standards Institution cannot be tolerated. Also, governors of the all-speed type, rather than the idling-and-maximum variety most used on road vehicles, are to be found on industrial versions of automotive units.

Because operating conditions frequently expose tractor and industrial engines to extremes of weather, ease of starting from cold assumes considerable

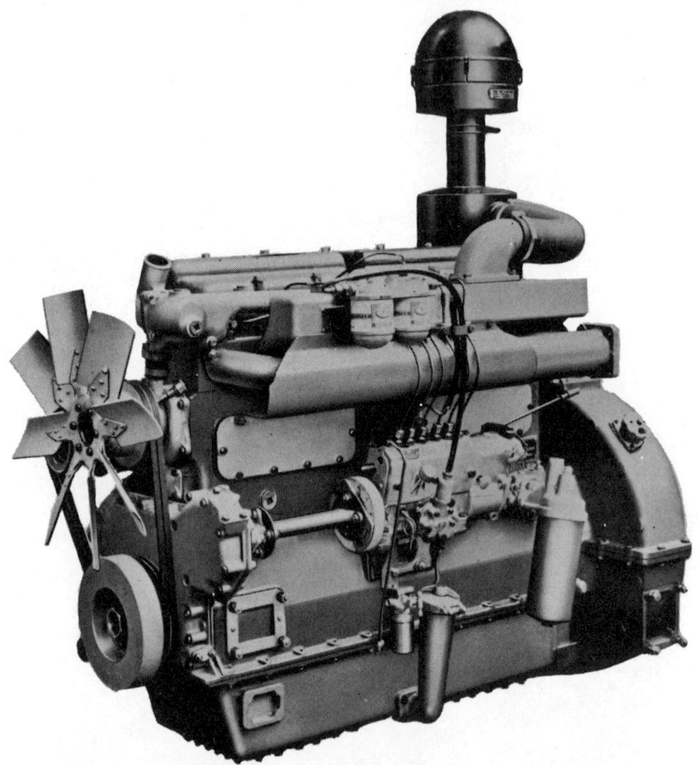

An industrial version of the 11·1 l Leyland 680 engine with pusher fan, none of the usual automotive auxiliaries, a full-length sump, vertical air intake, and large flywheel

importance. In this respect automotive diesels with direct-injection combustion have been able to score a distinct advantage over other types of engine and over some traditional types of industrial diesel.

Electric starting is still the normal equipment for these fast-running industrial engines, but options of various manually energised starters are also now usually offered. These special starters take the form either of an inertia mechanism or a hydraulic motor. In the first case, energy is imparted to a freely rotating flywheel, which is then coupled through suitable gearing to the

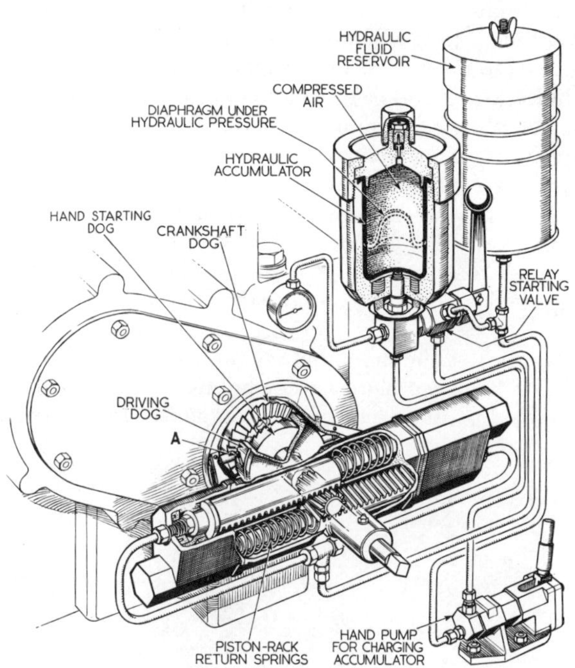

Bryce Berger Handraulic starter. The piston racks of a two-cylinder hydraulic motor rotate a ratchet device concentric with the starting handle

Inserting capsule in the CAV ether starting device. Screwing home the holder pierces the plastic bulb and releases the fluid

engine. With the hydraulic starter, pressure is built up by a hand-pump in a hydraulic accumulator, the energy then being applied to a hydraulic ram mechanism pushing a rack which meshes with a pinion on the crankshaft.

A factor leading to the increasing acceptance of diesel engines for industrial use is their ability to respond to full load very soon after starting, without the hesitation typical of engines running on petrol or vaporising oil. This attribute of diesel engines has been enhanced for several industrial applications by introducing an oil-to-water heat exchanger in the system so that the lubricating oil will gain its normal viscosity quickly. The heat exchanger also allows heat to be transferred from the oil when it becomes hot, with the result that the whole engine tends to remain at a constant and evenly distributed temperature, avoiding breakdown of the oil film and keeping the operating temperature at a stable figure.

The wide adoption of the small and medium-sized diesel engine in agricultural applications has been helped tremendously by the relatively small cost which is now added by diesel-powered equipment. This has reduced the time

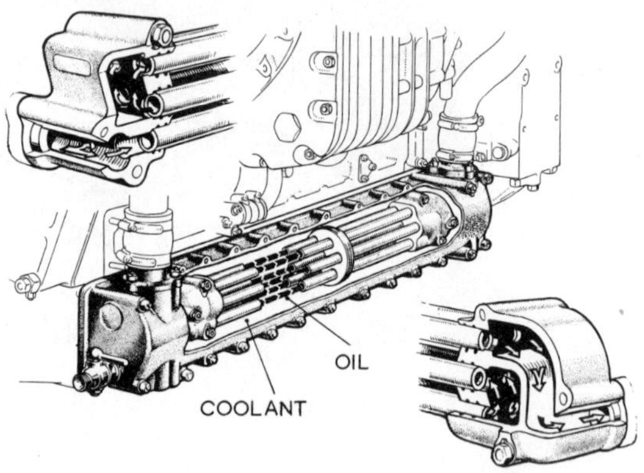

Heat exchanger on Rolls-Royce engines for rapid warm-up of lubricating oil by the coolant

over which a diesel needs to be run before it can offset its extra capital cost through its better fuel consumption. Most of the popular types of small agricultural tractors are now available with diesel engines, usually in the range between 30 and 70 bhp.

Under the more constant-load and constant-temperature conditions typical of industrial applications, automotive diesel engines give an even higher reliability factor than they do when installed in road vehicles. The only snag which can occur with industrial applications is the lack of understanding on the part of personnel responsible for using them. For example, it is not often appreciated how important it is to use clean fuel. Dirty petrol or vaporising oil does relatively little harm to carburetted engines. If the particles are large enough, they will either choke the filter or block a jet and the engine will stop, so forcing attention on itself. In a diesel fuel injection system dirt particles in the fuel will rarely be sufficiently big to stop the flow but, what is more serious, abrasive matter is introduced into the fine working clearances of the injection equipment. Enclosure of engines and even separate enclosure of injection pumps helps on this score but the greater danger of fuel contamination is by careless filling from dirty cans or funnels and neglect of fuel filters.

This Hungarian Dutra four-wheel drive tractor is fitted with a 35 bhp three-cylinder Perkins diesel

Export business in industrialised high-speed diesel engines has made big strides in recent years, especially in remote territories such as oil fields. The popularity of diesels for such applications is being further enhanced by the multi-fuel developments in diesels. Multi-fuel diesels can run without harm on whatever type of light-fraction oil product happens to be handy—from gas oil to high-octane petrol. In fact, for some applications, diesels run on liquefied petroleum gases (l.p.g.), such as propane. To run on l.p.g. requires special modifications, but fuel economy is still excellent, there is very little carbon formation and, what is interesting, diesel knock disappears.

12
Air-cooled diesels

For many purposes an air-cooled engine has considerable advantage, particularly in the small sizes where portability is a factor in which the bulkiness of a liquid cooling system is an embarrassment. In any case, the weight saved by the elimination of the usual radiator, the water-jacketing, and the coolant itself can be considerable, while the air-cooled unit is physically less vulnerable to rough usage. Also, the freedom from risk of frost damage is not to be underrated.

Apparent simplicity of installation has encouraged the development of small air-cooled engines for some time in industrial and certain marine propulsion applications, although it has sometimes been overlooked that in boats a large volume of fast-moving heated air has to be disposed of and that this can involve a considerable amount of air trunking. Something of the same problem accompanies the installation of an air-cooled engine in road vehicles, where some means has to be found of harnessing the engine's heat to serve the driving compartment by a method which is free from fumes. Nevertheless there is now more interest in air-cooled diesel engines than has been evident for some time.

Directly cooled engines need forced circulation provided by a multi-bladed fan driving air through a cowling enclosing the finned cylinder head and cylinder block assembly. The high thermal efficiency of the diesel is favourable to air-cooling in the sense that the total amount of heat to be dissipated is relatively low. On the other hand local temperatures can be very high and it is not easy, especially as far as air-cooled cylinder heads are concerned, to provide the extra cooling of the injector and valve seats which can usually be arranged with a water circulation system.

The basic simplicity of air cooling has long attracted road transport engineers and experimental work is continually going on. Few production engines have been seen for automotive use, however. From the Continent of Europe some of the commercially most successful automotive air-cooled diesel engines have emerged. These are the German Deutz and the Czechoslovak Tatra; they bear considerable visual similarity and are both particularly popular in their vee forms. Until comparatively recently they were all indirect-injection designs, but the more efficient direct-injection now makes the running. The Deutz is now made in V10 and V16 as well as V8 and V12 forms.

What is most difficult to arrange with air-cooled designs is that heat is conducted rapidly to the outside air before it has time to accumulate in local hot spots, so that one part of a component tries to expand faster than another part. This is especially acute in automotive use where the operating temperature is usually rapidly varying for much of the time. This is one reason why air-cooled engines have been more successful in marine and industrial applications, where the comparatively steady speed and steady-load conditions at conservative power ratings produce fairly constant temperatures, which are in any case typically lower than those reached in automotive installations.

Indeed, the successful use of air-cooled diesels on the industrial front has encouraged some manufacturers in this sphere to experiment with their engines in lorries and buses. Two British manufacturers of air-cooled industrial engines have been particularly active in this direction. They are Dorman and Ruston. For some years now Lincoln Corporation has run a bus powered by a Ruston air-cooled six-cylinder diesel. The same engine has also seen substantial

A Tatra in-line four-cylinder air-cooled diesel engine

service in an Atkinson articulated vehicle operated by one of the Transport Development Group haulage companies. A Dorman engine has been installed in a Ford D-series two-axle truck.

In all these cases, the basic reliability of the air-cooled engines has not given cause for concern. However, being based on industrial engine designs, the engines are heavy compared with automotive diesels of comparable power and neither is their power output as high as usual relative to their capacity. The development period is therefore inevitably going to be lengthy. If lighter, higher-revving and more powerful versions of these air-cooled diesels are

A Deutz V8 air-cooled diesel engine

Section through a Deutz air-cooled head and cylinder barrel

going to be made, they will have to be reassessed in their new form to see whether they can maintain their standards of reliability. As it takes about two years to run an engine the 200 000 miles' minimum expected of most British automotive diesels in these days, even with intensive round-the-clock operation, the answers will not be known for some time. By then rival liquid-cooled designs might have been developed to an even higher pitch, so that there would once more be odds against the air-cooled engine.

These commercial considerations make some manufacturers of liquid-cooled engines, already dominant as far as power output and mechanical

A Deutz air-cooled engine in a converted Guy double-decker bus operated by the Gosport and Fareham Omnibus Co. Ltd

reliability are concerned in relation to light weight, think they have a better chance of producing air-cooled designs of competitive status. In the 1950s, that bold bus operator which until recently manufactured its own engines and vehicles, Britain's Birmingham and Midland Motor Omnibus Company (better known as Midland Red), blazed the trail by making an air-cooled version of its $8\frac{1}{2}$ l. six-cylinder diesel. As it happened, the thermal stresses from the poor dissipation of heat from local hot-spots caused cracking problems. It became evident that the project would be too expensive for a small

Among recent experimental installations of air-cooled diesels in vehicles in Britain has been this Dorman 6·23 l. six-cylinder engine in a D-series Ford 7½ ton truck. This engine develops 107 bhp at 2 500 rev/min and development of automotive models is under way

A 3·93 l. four-cylinder air-cooled engine fitted in light trucks by Robur, the East German manufacturer. It develops 70 bhp at 2 600 rev/min

maker to develop, and so the idea was dropped. Unfortunately, this pioneering experience promoted extra caution among the big engine manufacturers and, although there is still plenty of discussion and experimental work behind the scenes, no British manufacturer already established in the automotive field has since seriously taken up the development of an air-cooled engine.

Despite the general pessimism there is still enough enthusiasm for air-cooled diesels among a few development engineers scattered throughout Britain and the USA. One of these, Mr. G. Lee, years ago got as far as building, when he was with Albion Motors Ltd., a single-cylinder diesel with unusual features designed to overcome the drawbacks of air-cooling. He had been closely associated with Bristol air-cooled radial aero-engines and applied some of the aero-engine techniques. Vertical instead of horizontal finning improved air flow and heat dissipation. Use of aluminium alloy suppressed noise and also relieved thermal stresses. A novel valve gear overcame the problem of

In Tatra V12 air-cooled diesels, air is blown first to the cylinder heads and then passes out between the cylinders

excessive tappet clearances when hot. Air was blown to the hottest parts (the cylinder head) of the engine first. This last feature applies to the Tatra.

Aluminium alloy does not find much favour yet among diesel engine makers, although more engineers are now recognising that modern alloys which are both stronger and more homogeneously cast are now available. Also, the attraction of aluminium alloys grows as engines become more powerful and bigger, because the weight saving then becomes more significant and more valuable. The thermal expansion rate can be embarrassing, but the very good heat conductivity of aluminium makes it a strong contender for future air-cooled designs. Of course it has to be admitted that much development would be needed on an engine built from a less traditional material, and the cost and time involved are formidable enough to make a manufacturer think twice.

The noise level of an air-cooled diesel, as far as experience shows up to now, is undoubtedly higher than that of a liquid-cooled engine, although little work has been done on designing finning so that sound attenuation is damped. In a world becoming increasingly sensitive to noise from mechanical contrivances, this aspect is more serious than might appear at first sight. Shrouding can muffle engine noise, of course, but this puts up the installation bulk, cost and weight again—offsetting the advantages which an air-cooled engine should offer. It is therefore essential, from an economic point of view, to suppress noise emission at its source.

Noisiness in air-cooled diesels is not helped by the greater internal expansion which results from the higher working temperatures which are usual. Excessive clearances tend to develop in working parts, affecting noise, power output and oil consumption. Hotter running also places heavier demands on the lubricating oil, although this is in any case becoming a familiar situation on turbocharged and two-stroke diesels as well, and is being overcome.

All these costly development problems with automotive air-cooled engines seem likely to mean that existing established designs will remain pre-eminent in air-cooled diesel practice for a considerable time, despite some inherent shortcomings which are so far perfectly tolerable in service. For industrial applications, however, air cooling could gain ground faster and it would seem to be on the issue of cooling methods that the ways of automotive engine design and industrial engine design begin to part.

13
Marine service

In recent years there has been a revolutionary change in the attitude of the marine world to the automotive diesel. It was the exhibition in 1929 of the Gardner engine in Britain and the Cummins in America that called attention to the possibilities of faster running and more compact engines as propulsion units for small vessels in which the power requirement from a single engine was in the 50–100 bhp range. Of course, the original 'lightweight high-speed' engines were only so designated in terms of the relative values of their period. An engine such as the Gardner L2 type, operating at 1 300 rev/min and having a specific weight of some 26 lb/bhp was indeed 'high-speed' when compared with the generally accepted 800 rev/min, and it was 'lightweight' when the prevailing figure was more often 50–70 lb/bhp.

Subsequent development was almost wholly the result of these pioneer lightweight high-speed marine diesels being adapted for road transport service, whereby they were brought into the sphere of quantity production in which engines and components could be considered in thousands as against the single installation or small-batch system that prevailed in the marine engine field. As soon as the power units of road transport vehicles were successfully established their possibilities in marine application were eagerly examined.

Quantity production techniques were in action, essential development work had been completed at the expense of an industry more progressive and enterprising than that associated with boat and yacht building and, again through the organisation and facilities of the road vehicle industry, there was an already established world-wide off-the-shelf service of spare parts.

Most leading automotive-type diesels are now available in marine form, the basic engine unit being equipped with suitable external components such as coolant and bilge pumps, heat exchanger fresh-water cooling systems, marine gearboxes and other modifications necessary for propulsion purposes on the water. The basic automotive unit provides a neat, compact and externally 'clean' engine and offers high performance with good reliability. First cost is much less than that of an engine of equivalent size made on the batch production basis, while spare parts are cheap, interchangeable and obtainable wherever the engine makers have a service depot for road vehicles. An important point too is that mechanics skilled in engine and injection equipment work are always to be found where diesel trucks and buses are operated with engines of the same design.

Acceptance of the automotive-type diesel in pleasure craft and in work boats of many types has become almost universal, so much so that many makers of the traditional low-speed diesel, although they have never been associated with the automotive industry, have now developed engine types clearly influenced by the design and performance characteristics of the transport diesels.

When road transport and marine engines are compared, it must not be overlooked that certain differences in performance requirements are involved and these result in variations in the basic specifications. For example, the road vehicle engine is running a great deal of its time under part-load conditions with only intermittent demand for maximum power; but high acceleration rate is essential. In marine propulsion, on the other hand, full load must be held for long periods while high speed, as such, is not a factor of importance, nor is acceleration. For most work, therefore, the engine is de-rated from its automotive settings. Governed speed and fuel pump delivery are often reduced by some 10% or more with corresponding reduction of power output. An engine with a maximum automotive rating of 120 bhp at 1 800 rev/min may have a marine rating of 100 bhp at 1 600 rev/min, the difference being that the former is based on a 1 hour demand whereas the latter will be guaranteed to be held for 12 hours or alternatively may even be a 'continuous' rating. For racing, however, the pattern is completely opposite; automotive engines are then up-rated and supercharged so that they commonly achieve a specific weight of only 6 lb/bhp.

By de-rating an engine its specific weight becomes correspondingly less favourable and the 10 to 14 lb/bhp of the automotive unit may become 16 to 17 lb/bhp in the marine form (for the engine without accessories). Weight, of

SPECIFIC WEIGHT (lb/bhp)

Make	*Automotive*	*Marine*
AEC	12·5	25
Ailsa Craig	—	35
Gardner	14	27·5
Leyland	14	23
R.N.	—	42
Ruston	—	52

Substantially equivalent types compared. It is evident that road transport development has been mainly responsible for weight reduction in marine units

course, is not regarded with quite the same disfavour in a vessel as it is in a road vehicle and direct comparisons do not serve a very useful purpose. Such comparisons are indeed difficult to make, because it is usual to quote the weight of marine engines complete with cooling equipment, reverse gear and other essential marine components. This may be regarded as a trade custom covering the entire 'machinery' rather than as a technical assessment of the power developed from a given weight of engine.

As matters stand, the reduced weight of many installations today reflects the great influence of the automotive types on marine propulsion units within their size range. The automotive-type marine unit has about double the specific weight of its road transport prototype, while those engines of purely marine origins and without road transport associations may again be double the average of the 'automotive-marine' type. For power-boat racing, automotive-derived diesels give even more power than they do in their vehicle applications. Full advantage can be taken of the copious cooling possible with marine work and this enables very high power outputs to be obtained—and sustained for long periods. The big success of Ford, Perkins and Foden turbocharged diesels in power-boat racing points to the large power potential of modern diesels. Cummins V8 and Leyland fixed-head engines have also been pushed to high powers on such marine work.

An important aspect of the automotive engine applied to marine work is the universal application of closed-circuit fresh-water cooling by means of a heat exchanger fed by sea water. This has been adopted as a result of the intensive research into causes of cylinder bore wear and piston ring trouble. Overcooling or slow warming up were discovered to be at the root of the matter,

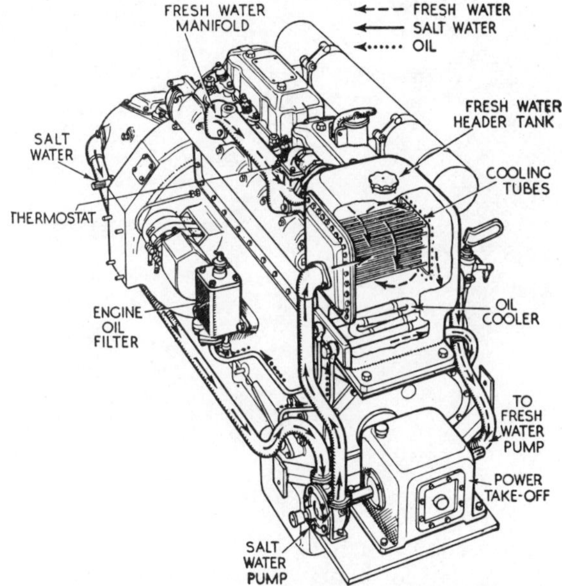

In a fresh-water closed-circuit cooling system sea water is pumped through a heat-exchanger which replaces the normal automotive air-cooled radiator. A heat-exchanger unit for the lubricating oil is also incorporated. In this example gear-cooling is effected by leading the sea-water intake through a water jacket

and nothing was more conducive to these conditions than the 'once-through' type of sea-water cooling commonly used on small marine engines of traditional type, particularly on fishing and other work boats. An objection to closed-circuit cooling is that two pumps are necessary, the normal circulation pump on the engine and a sea-water pump to supply the heat exchanger.

The latter can be avoided by circulating the fresh-water coolant through tubes laid externally along the keel; such an arrangement is vulnerable, however.

Fuel consumption is another matter on which there is not the same approach. In marine service a corresponding figure to road vehicle mile/gal is not readily available; the most useful statement of consumption is the rate at governed speed in terms of gal/h. Engine makers quite often prefer to quote the optimum test-bed figure in pt/bhp h which is of undoubted academic interest but is likely to involve an owner in some frantic mental arithmetic if, in the midst of a passage, some uneasiness arises as to the number of operating hours still available in the fuel tanks. In such circumstances pints (or gallons) per hour provide the operating engineer with a far more acceptable basis for rapid assessment of the position than do the pt/bhp h of the test-house technician.

The next consideration is that of fire risk. A fire on board ship is infinitely more dangerous than one on shore. The fuels used in oil engines do not readily evaporate in ordinary atmospheric conditions; consequently there is no invisible and violently sudden danger attendant upon a leak in the fuel system. The absence of a high-tension electrical ignition system likewise precludes the risk of stray sparks capable of igniting petrol or petrol vapour. That the reduction of fire risk is very real is reflected in the lower insurance rates

Cummins V8-300 marine engine pack with integral cooling system

obtainable in respect of diesel-engined vessels; there is no specific percentage reduction but it is general for underwriters to quote lower premiums than those applicable to equivalent petrol installations.

As regards performance, conditions afloat are in no way as variable as those on shore. The average marine engine is almost a constant-speed unit. It is designed to operate continuously at its predetermined governed power output for any reasonable length of time. A spell of 100 h should be well within its scope.

A certain amount of flexibility is required, but this is usually interpreted to mean an ability to turn over very slowly under the control of the governor and accept without hesitation the engagement of either the ahead or astern clutches without reference to the hand-throttle control.

At this kind of flexibility the oil engine excels as the governor controls the amount of fuel injected. In other words the governor takes care of variations in torque loading and the 'throttle' is purely a speed control. With a petrol carburettor violent opening of the throttle will sometimes lead to stalling, but rarely so with a diesel engine. It will thus be realised that in marine service the engine is generally operating well within its safe limits. In construction the compression-ignition engine must of necessity be of robust design and by virtue of its performance characteristics also it is predisposed towards reliability. In most cases crankshafts exceed Lloyd's requirements for open sea service.

Certainty of starting is an essential requirement in a marine engine. In this respect the diesel has most favourable claims to consideration. The direct-injection type, particularly those with the simplest form of open-cavity piston

The Fairey Sea Fox powerboat is driven by two turbocharged Perkins T6.354 six-cylinder diesels

and multi-hole injectors, are remarkably easy to start from cold, even in sub-normal temperatures. Hand starting of six-cylinder engines of about $4\frac{1}{2}$ in cylinder bore is indeed possible, although most modern installations have electric starters. Ability to start by hand is not to be dismissed as an out-of-date characteristic, however, because batteries may run down during periods of neglect, and there may be no readily available means for recharging them from an external supply.

Large engines with air cell combustion chamber systems can rarely be hand-started from cold unless heater plugs are fitted in the cylinders or a heater coil is mounted within the air-intake manifold.

Provision in some cases must also be made for an initial injection of kerosene while for extremely low-temperature starting an apparatus for admitting ether may have to be installed.

When heat must be applied electrically this increases still further the demand upon the battery and although certain types of engines will start readily with 12 V starter motors it has become usual on automotive type units to incorporate 24 V electrical equipment so that the engine can be spun at fairly high speed, thus not allowing time for heat loss from the combustion chamber during the compression stroke.

At one period compressed-air starter motors had some following; there are also hydraulic motors and even explosive cartridges which admit high-pressure gas into one of the engine cylinders. However, these may be regarded as expedients for special duties and conditions; the electric starting system has attained a high degree of popularity because it works readily, involves no effort and, providing the batteries are not allowed to run down, is very reliable. In marine conditions there is no doubt that only the very smallest engines should be started by hand. It is so often nearly impossible to exert full power on the handle when the engine is installed, since, in many cases, its location and the movement of the vessel combine to make matters difficult.

No special progress has been made in the British marine field with the two-stroke multi-cylinder engine in the 70–130 bhp range. This is a further

A Perkins 4.107 stand-by marine diesel in the yacht Gipsy Moth IV

indication of the dependence of the light marine engine industry on automotive development. In so far as marine propulsion involves constant-load, constant-speed conditions, it should be favourable to the two-stroke system, whereas the characteristics required in automotive usage are quite different, involving a widely fluctuating speed and load range, with demand for high accelerative performance.

It remains to be seen if the situation will be influenced by the impressive adaptation of the opposed-piston two-stroke embodied in the latest 2 500 bhp lightweight units first developed for high-speed naval craft and more recently applied to new types of railway locomotives. The possibility certainly arises, although it may be pointed out that the basic engine type has been available in various automotive, marine and aircraft forms for 30 years without securing an established position in any of them. Almost identical comment might be applied in the case of the 3·6 l. three-cylinder opposed piston two-stroke which has become a fairly prominent unit in the road vehicle field and which also is available in marine and industrial forms. However, it is not safe to draw conclusions in respect of engines of such widely different power outputs. Whereas the two-stroke has never received unqualified acceptance within the size range of the automotive field, it has been highly successful in very large units operating under steady conditions. But there are two notable exceptions, the well-established and successful two-stroke marine units which are counterparts of automotive engines of equivalent design, the American GMC and the British Foden. The Foden two-stroke is a favourite with the British Admiralty.

The two-stroke engine necessarily requires forced induction from a suitable blower unit and all the engines named have positively driven scavenge blowers of the Roots type. Supercharging can, of course, be applied to four-stroke engines to increase their power output but the four-stroke unsupercharged engine is still the predominant type, and in the performance range of 4·5–12 l. capacity (300–700 in^3 approximately), developing from 40 to 180 bhp at continuous rating, the six-cylinder arrangement is the most popular. Four-cylinder engines predominate, however, in the smaller class of 4 l. capacity or under.

Engine speed appears to be stabilised at about 1 600 rev/min in the case of engines normally operated intermittently at 1 800 to 2 000 rev/min in automotive service. Fuel consumption of about 5 gal/h may be considered normal for a six-cylinder engine developing 100 bhp, while engine life between overhauls may be expected to be not less than 10 000 hours for the best automotive types. For higher powers the trend is towards turbocharged six-cylinder and unsupercharged V8 engines (now available up to 14 l. capacity).

Apart from its application to propulsion machinery, the small high-speed diesel has considerably influenced the design and layout of auxiliary equipment on larger vessels. Many automotive engines have now been adapted for industrial use and are embodied in self-contained direct-coupled generator sets, pumping units and compressors, and they are being applied to many auxiliary uses in the engine room. Even on relatively small craft it is still possible to install a separate electric light and battery charging plant by using one of the many small single- or twin-cylinder air- or water-cooled diesels developed in recent years. These engines for stationary duties are referred to in the previous chapter.

There is a fundamental matter to be considered when high-speed engines are to be installed in pairs; does the design and duty of the craft require units of opposite hand—one rotating clockwise and the other anti-clockwise.

It is most important to ensure that the engines are properly marinised types, built and equipped to withstand the inevitably damp conditions of service, whether on salt water or fresh. Diesel engines, however, are not so subject to hygroscopic troubles as are carburettor engines, with their electric ignition apparatus.

When being installed it is essential to remember that fuel and oil filters, and other items requiring regular service attention, should be accessible from the top of the unit, as that is the location most commonly convenient to the operator; this is especially important when the engines are mounted low in the hull.

Particularly in the case of passenger-carrying boats, as distinct from work boats, a valuable installation feature is proper acoustic treatment, including such features as resilient mountings and acoustic 'hooding' to minimise transmission of vibration and noise. In conjunction with noise it must be borne in mind that the operator handling the boat may be out of earshot of the engine, to a far greater extent than in the case of a vehicle power unit. Simple instrumentation, such as a tachometer, can advantageously be provided to help the user handle the engine to the best effect, especially when manoeuvring.

One condition which must be regarded is the fact that many boats are apt to be idle for long periods, especially when used for pleasure purposes. This affects not only equipment such as starter systems, filters and ease of operating the starting handle on smaller units, but choice of supplies such as lubricants.

14
Diesels in cars

Whether diesel engines find application in some particular field of motive power nearly always depends on the extent of the economic benefits they can bring. Those benefits are less clear when it comes to powering ordinary cars. Certainly the diesel engine, as usual, shows greatly improved fuel economy compared with petrol engines of either equivalent size or equivalent power. For example, an improvement in miles per gallon from 28 to 40 is often achieved (or 45 to 60 km on 4·5 l.).

But impressive as is the gain in fuel economy, expressed as a percentage, it has to be remembered that the actual volume of saved fuel represented is much smaller than is the case when the fuel consumption of a heavy vehicle is improved from 5 to 8 mile/gal (8 to 12·8 km on 4·5 l.). It follows that the actual cash value of the fuel saving is greatly diminished in the case of cars.

Moreover, in Britain at least, the middle-distillate fuels, of which diesel oil is a major member, are getting in short supply as the pattern of demand for oil products has progressively departed from the usual distribution of output from refineries. Heavy demand for industrial fuel oils, which are heavier than diesel (or gas) oil, has automatically produced something of an excess of the light fractions, which to all intents and purposes means petrol. The old commercial laws of supply and demand have come into play so that diesel fuel has increased in price while the price of petrol has been cut. Thus, the situation in Britain is now that diesel fuel costs, retail, as much as 98-octane petrol. Consequently, the diesel's advantage of using a cheaper fuel has been eliminated.

This state of affairs has not been reached in many other countries, however—a point often not sufficiently appreciated by most British companies concerned with diesel-engined cars. In the Middle East and Eastern Europe, for example, diesel fuel is still comparatively cheap.

Even where diesel engines have the attraction of using a cheap fuel, however, their applications are still mainly to cars subject to intensive usage. And this immediately means taxis. In the Middle East, Belgium and Portugal, big American cars can often be seen used as taxis, but fitted with Perkins four-cylinder engines to ensure the lowest possible running costs. In fact, as far as diesel engine manufacturers are concerned, the British Perkins company has probably done more than any other in the car conversion field.

It is believed that the first taxi diesel engine was born in 1934 when a 45 bhp Perkins Wolf engine was fitted in a Studebaker car. Since that date a wide variety of cars all over the world have been fitted with a Perkins engine at one time or another.

In 1938 diesel taxis with Perkins P4 engines were operating in London, Manchester and Edinburgh. In 1939 war halted the plans of Nash, of America, to introduce a Perkins-powered taxi, but after the war both Belgian and Portuguese taxi fleet operators turned to Perkins for hundreds of engines. A Brussels company purchased 350 Chevrolets fitted with P4 units; a similar

Installation of a Perkins 4.108 diesel in a Hillman car

number went into service in Antwerp. In Portugal about 500 taxis were fitted with Perkins four- and six-cylinder engines in both American and European cars.

So successful was the Perkins design that for some time it was offered as first equipment in the Plymouth range of cars produced by Chrysler's Antwerp company. Fuel consumptions of many Russian Zim, Pobeda and Volga cars, used as taxis in Finnish towns and cities, were approximately halved by the installation of these Perkins four-cylinder engines.

There came a growing interest in automotive diesels in the United States where rising costs prompted both transport and cab operators to study the successful operation of diesel-powered fleets in Europe and elsewhere. The Perkins company's engines have already been installed in taxis in a number of US cities including New York, Detroit, Los Angeles, San Francisco, Seattle, Boston, Atlanta, Cleveland and Trenton. It is estimated that in the early 1960s several thousand Perkins-engined taxis were in use in the United States.

When Perkins introduced its 1·6 l. 4.99 diesel engine in 1961 much enthusiasm was aroused among taxi operators and car fleet owners throughout the world. The 48 bhp 4.99 high-speed engine was soon offered as initial equipment in British Beardmore taxis. In Finland the unit was installed as first equipment in Russian Volga cars. The 4.99 was also fitted in another Russian car, the Scaldia Moskvitch, by a Belgian company.

More than 1 500 Perkins 4.99 engines were ordered for Malaya and Singapore in 18 months and at least 95% of these were for installation in taxis, including Ford Consuls and Zephyrs, Morris Oxfords, Austin A55s, the Hillman Minx and the Australian-built Holdens.

Among the first fleet operators to convert comparatively large numbers of cars from petrol to the 4.99 were Perkins themselves. The conversions were carried out to 18 Vauxhall Velox and Wyverns and four Ford Consuls. The Hillman Minx was also fitted with the 4.99 and returned 57·9 mile/gal at a 30 mile/h average, although 44 mile/gal was a more usual overall figure (93 km on 4·5 l. at 48 km/h average, but 70 km is more usual). Conversion kits were made available through Perkins for Ford Zephyr-Four, Morris Oxford, Hillman Minx and other popular models.

Another Perkins engine popular in large American taxis both in the United States and in Europe is the 3·33 l. four-cylinder 4.203 unit.

In 1965 Perkins announced another four-cylinder diesel engine designed specifically to meet the high power requirements of modern light commercial

The 1·76 l. Perkins 4.108 engine

vehicles, taxis and cars. The new unit, the 1·76 l. 52 bhp 4.108 engine, developed its maximum power at 4 000 rev/min—a speed once considered impossibly fast for a diesel engine. At the International Poznan Fair in June 1967 a Polish-made Warszawa 203 five-seater car was exhibited for the first time with a Perkins 4.108 engine as standard equipment.

Yet it is bound to be the car manufacturers themselves who will always be in the best position to offer diesel-engined cars at a competitive price. Conversions are relatively expensive, but the fact that so many operators of high-mileage cars throughout the world find conversions economically worth while rather proves that here is an important market not catered for as widely as it might.

There is an outstanding exception to the general indifference of car makers to the diesel engine, and that is the German Daimler-Benz company, whose Mercedes-Benz 180D, 190D, 200D and now 220D diesel-engined cars can be seen (used mainly as taxis) all over the world. Mercedes-Benz diesel cars recently topped the million mark.

The Austin-Morris division of the British Leyland Motor Corporation has also had great success with diesel-engined taxis. But in this case world-wide

The Polish Warszawa car is powered by a four-cylinder 1·76 l. Perkins engine developing 52 bhp at 4 000 rev/min

extension of the success of its 2·2 l. engined British taxis has been strangled by the lack of appeal that the London taxi design as a whole (dictated by London's Metropolitan Police) has for any other sphere of operation.

To some extent, the Austin management recognised the need for a more conventional diesel-engined car for markets other than British cities by introducing a diesel-engined version of the A60 car. This was powered by a diesel version of the company's 1½ l. petrol engine and it has proved quite successful. What is more, the extra price of the diesel-engined A60 car was only £70, which is better than has been achieved by most other manufacturers, although Fiat, in Italy, and Peugeot, in France, are also now tackling the small diesel car-market seriously. Even so, the economics are such that a car needs to do about 40 000 miles (64 360 km) a year for any significant benefit to show in the operating costs.

Experiments in applying diesel engines to cars have long been made, and will no doubt continue to be made. These are of considerable academic interest and can even yield much useful information for the further development of high-speed diesel engines, but they are hardly ever an economic proposition. When the late Mr. G. Waring was research engineer of Leyland Motors Ltd. he installed a six-cylinder 350 in^3 (5·7 l.) diesel engine in a large old Bentley saloon. The experiment taught the Leyland engineers a lot in engine tuning and in a diesel's behaviour at high rev/min.

Another Lancashire company, L. Gardner and Sons Ltd., has installed its diesel engines in cars occasionally for many years. The latest is a Jaguar XK150 sports car powered by a 3·8 l. Gardner 4LK engine running ungoverned and developing more power than was ever dreamed of from this engine. It regularly returns over 45 mile/gal at high average speeds, and it will exceed 100 mile/h. Apart from teaching much about high-speed power, this engine installation has brought great benefits in mounting technique. But the 4LK is not a small diesel engine by present-day standards and would be too expensive to be considered for regular fitment in cars.

More development needs to be done on small diesel engines before they can hope to gain a much bigger slice of the car market. Designers need to attack on four fronts: cost, power output, noise, and efficiency.

The higher price of diesel engines stems principally from the need for heavier castings and from the inherent costliness of a fuel-injection system compared with the petrol engine's carburettor and ignition system. But the

A BLMC 1½ l. diesel engine with car transmission bolted to it

price differential has been narrowed in recent years by the advent of successful distributor pumps, higher production runs and stronger metals for castings. Sharing components with an existing quantity-produced petrol engine also helps in reducing the cost gap.

The price set on small diesel engines therefore becomes more and more a matter of what the market finds worth while to pay rather than what the intrinsic extra cost actually is. Nevertheless, a price differential against the diesel will probably be unavoidable always and, in any case, a differential less than £60 in retail-price terms is unlikely to be achieved.

Power output from small diesel engines is limited. It is limited by exhaust smoke density, by frequent restricted ability to run much above 4 500 rev/min and by a restricted valve timing which spoils volumetric efficiency at high speeds. Until a much better specific power output is obtained from small diesel engines they cannot make a serious impression on the ordinary car owners. Expensive development is needed on small diesel engines of 2½ l. and less.

The ways of obtaining a higher power output might have to break fresh ground. For example, the reason a diesel's valve timing is restricted is because the clearance between cylinder head and piston crown is extremely small when the piston is at the top of its stroke. Valve overlap is forced to be small for

Specially tuned version of the 3·8 l. Gardner 4LK engine for car installation

A Gardner 4LK fitted in a Jaguar XK150 sports car

fear of the valve heads touching the piston before it has been able to retreat sufficiently down the bore. This especially applies to the valve-timing overlap at the changeover from the exhaust to the induction stroke; when the valves cannot be kept open as long as would be liked in this period no use can be made of inter-extraction effects between the exhaust gases rushing out and the fresh intake of air being drawn in.

Even if just the exhaust valve could be kept open later near the top of the exhaust scavenging stroke it would enhance power output and thermal efficiency. To do this, however, the exhaust valve head would have to be recessed so that extra clearance was provided between valve head and piston.

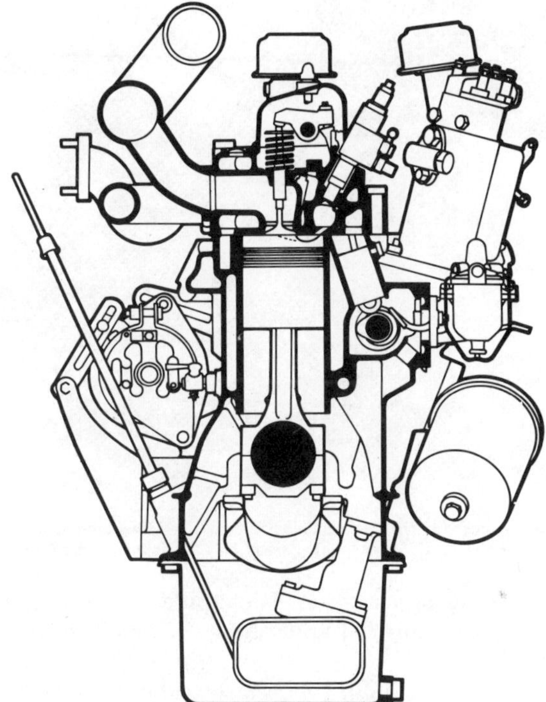

Cross-section through the Rover 2¼ l. diesel which has Ricardo Mk V indirect-injection

Such recesses normally spoil combustion efficiency, but it is not beyond the bounds of possibility that either the whole of the combustion chamber, or a secondary chamber linked with an air cell, could be concentrated beneath the recessed exhaust valve. The fuel injector could then be nearly horizontal and be accessible on the outside of the cylinder block. The outcome could well be a combustion chamber virtually a cross-breed between an indirect-injection air cell and a direct-injection piston cavity. As with an air cell, practically all the air swirl would be as a result of squish and there would be freedom to design the porting almost entirely from the point of view of getting the easiest entry and exit of gases. There would be a lot of wetting of the combustion chamber with fuel, as with an air cell or the M.A.N. Meurer spherical piston cavity, so that the pressure build-up would not be so violent and noise would be reduced to some extent.

Fresh approaches such as this are now being considered by some designers. In the meantime, the highest specific power output from small diesels is given by the Ricardo Mk V indirect-injection layout. Direct-injection has been tried more than once by Leyland and Standard–Triumph engineers in an effort to gain the greater efficiency, easier cold-starting, and simpler maintenance of direct-injection. Conventional direct-injection adaptations have never been able to run as fast as the indirect-injection versions, however, and

A Peugeot 1 255 cm³ four-cylinder diesel for mounting transversely in a front-wheel drive car; it develops 45 bhp at 5 000 rev/min. Peaking at 5 450 rev/min, this is one of the fastest running small diesels

have not developed as much power. Direct-injection of pure form therefore seems more suited to industrial applications of small diesels rather than to automotive installations.

Some way ought to be found, none the less, of improving the higher-speed efficiency of small diesel engines. At present their specific fuel consumption is not impressive and at full load can even be almost equalled by a very good petrol engine. If the small diesel's fuel economy advantage could be made even more emphatic there would be more users prepared to tolerate its drawbacks.

Diesel engines therefore seem destined to comprise only a minor proportion of car power plants for some years to come. But they are far more popular than only ten years ago and they find ready justification in economic terms for cars doing large annual mileages.

15
Automotive engine review

AEC (AEC Ltd., Southall, Middlesex, England) (part of British Leyland Motor Corporation)

Diesel engines were developed by AEC in 1930 for London buses. The first six-cylinder engines had Acro air cells, but another form of indirect-injection, the Ricardo spherical chamber, was used before long. These early engines were made in 6·6 and 7·6 l. forms. They were among the first to use lead/bronze bearings as a means of reducing bearing width and therefore engine length.

From these engines was derived the 7·7, an engine of 7·58 l. capacity, first with indirect-injection, then with direct-injection into a plain pot cavity in the piston and eventually with a deep straight-sided toroidal cavity, used right up to the present day by AEC. The 7·7 developed 98 bhp at 1 800 rev/min and was in production for 12 years. It was finally superseded after the war by a 9·6 l. engine originally designed in 1939 for railcars and London buses. This had timing drive by gear instead of chain and the camshaft in the crankcase instead of high in the block. It developed 125 bhp at 1 800 rev/min in its original form. A larger bore, 11·3 l. version followed later, this developing 150 bhp.

Subsequent AEC in-line engines have been developments of these 9·6 and 11·3 l. units. Wet liner editions, known as the A590 and A690, governed at 2 000 rev/min followed. In the same era, scaled down versions known as the A470 and A410 (7·68 l. and 6·75 l.) were introduced, and for a short time there was a four-cylinder version of these as well.

In 1966 the AEC range of engines was rationalised to some degree and a reversion made to dry liners. At the same time two new larger bore models were introduced to extend the power range. These were the A505 of 8·2 l. (descended from the A470) and the A760 of 12·47 l. (descended from the 9·6). Powers of the other engines were stepped up. There are now the 8·2 l. A505 (151 BS bhp at 2 400 rev/min), the 11·31 l. A691 (200 bhp at 2 200 rev/min), and the 12·47 l. A760 (220 bhp at 2 200 rev/min) in the AEC automotive range. Industrial versions of all these in-line six-cylinder engines are made. A minimum specific fuel consumption of 0·36 lb/bhp h (163 g/bhp h) is obtained; the on-road fuel consumption is among the better of the world's diesels.

For extra heavy-duty work such as earth-moving and industrial applications, both normally aspirated and turbocharged models of a 17·75 l. six-cylinder engine are produced. Both the bore and the stroke of these engines are 6·125 in (156 mm). Normally aspirated, the BS bhp is 289 at 1 900 rev/min. When turbocharged the power is 377 bhp. Corresponding maximum torques are 930 and 1 230 lbf ft (128·9 and 170·4 kgf m). They are similar to the smaller engines but have wet liners and four valves per cylinder.

AEC's latest oversquare V8 offers greater power without extra torque, which would necessitate heavier and more expensive transmission. With a stroke of only 4·5 in (115 mm), at 2 600 rev/min piston speed is well within 2 000 ft/min (10·1 m/sec). The 5·12 in bore (130 mm), 12·15 l. V8 produces 247 bhp (BS) or 265 bhp (SAE). Its maximum torque is 580 lbf ft (80·1 kgf m) and its minimum specific fuel consumption is 0·369 lb/bhp h (167·3 g/bhp h). Five-hole injectors are used. Cast iron predominates in the construction, but the weight is not much more than that of the 12·47 l. 760 in-line six. The AEC V8, known as the Leyland 800 V8, has been adopted as one of the new-generation power units for the British Leyland Motor Corporation truck division. A bigger-bore (135 mm or 5·31 in) version of 13·1 l. capacity is the latest product. Known as the 801 V8, this develops 272 bhp (BS) and a maximum torque of 638 lbf ft (88·1 kgf m) at 1 400 rev/min.

BEDFORD (Vauxhall Motors Ltd., Luton, Bedfordshire, England)

Although a subsidiary of the American General Motors Corporation, which has a well-established range of two-stroke diesels, Vauxhall made out a strong case for developing its own European-style four-stroke diesel engine, when it came to deciding to launch into diesel manufacture in Britain. The decision to make a four-stroke of independent design has paid off handsomely. Production of the first Bedford six-cylinder direct-injection diesel began in 1957. It was a 300 in^3 (4·927 l.) unit, called the 300, developing 97 bhp net at 2 800 rev/min, and it is still the basis for present Bedford engine designs.

Demand for more power inspired the introduction of a bigger-bore version of the 300, hence the 330 was born. It is one of the four Bedford diesels in the present range. Its capacity is 330 in^3 (5·42 l.) and it develops 107 bhp (or 112 on SAE rating) at 2 800 rev/min. A four-cylinder derivative, called the 220, with the same 4·063 in (103·1 mm) bore and 4·23 in (107·4 mm) stroke, is made. This has an in-line fuel injection pump whereas a CAV distributor pump is fitted to the six-cylinder 330.

In 1966 two bigger six-cylinder Bedford engines were introduced when heavier trucks were launched. Originally termed the 60 and the 70, they have many common features, being basically the same engine. With both models the stroke is 4·75 in (120·7 mm), but the 60 has a 4·125 in (104·8 mm) bore, giving 381 in^3 (6·243 l.), and the 70 has a 4·563 in (115·9 mm) bore, making a capacity of 466 in^3 (7·6 l.). The governed speed is 2 800 rev/min, at which the 60 develops 115 bhp and the 70, 135 bhp. A distributor fuel pump is fitted to the 60, but an in-line pump to the 70.

Within a generally normal arrangement, Bedford engines have some noteworthy features. After a lot of experimental work it was decided to take the bold step of not having any cylinder liners. This is said to improve cooling and eliminate stresses in the cylinder block casting, as well as saving cost. A steel gasket is used, and this has the sole function of sealing against gas pressure. There are no communicating water passages between block and head passing through the gasket; it can therefore be designed expressly for that one purpose. Water transfer between block and head takes place through a hollow elbow at the back of the engine, so that water flow is from end to end of the engine.

The valves are aluminised to make them more resistant to burning and corrosive action. The cylinder head studs are disposed so that six encircle each bore, and the top deck of the 60 and 70 cylinder blocks is 1 in (25·4 mm) thick to give greater freedom from distortion. Thermal stresses are minimised in

The Bedford 7·6 l. '70' engine

the head by alternating inlet and exhaust ports along the length. Three compression and two scraper rings are fitted to each piston. The offset toroidal combustion chambers have straight sides and an unusually shallow central pip.

In their latest forms, the 60 and 70 Bedford engines have been extensively modified to make them quieter. The cylinder block/crankcase has had several horizontal ribs added, the bottom flange is thicker, and the side walls (except in the region of the tappet gallery) have been thickened by 0·125 in (3·15 mm). These stiffening modifications have increased engine weight by 38 lb (17·2 kg). Fan noise has been reduced by irregular spacing of the blades, while the inlet manifold has been replaced with a new AC-Delco unit incorporating air filter elements. Piston noise has been minimised by a reduction of 0·002 in (0·05 mm) in bore clearance and by eliminating the skirt piston ring.

In the case of the 466 in^3 (7·6 l.) engine, the in-line injection pump has been superseded by a CAV distributor pump which has enabled maximum rev/min to be brought down from 2 800 to 2 500, yet with a net gain in maximum power of 3 bhp. At the same time, maximum torque is now developed at 1 000 rev/min instead of 1 600 rev/min.

BERLIET (Automobiles M. Berliet, Venissieux, Rhône, France)

Having, for many years, employed one form of antechamber combustion system or another, such as the Acro, Ricardo Whirlpool and Ricardo Comet Mark III, in its own-make engines, this French truck and bus manufacturer entered into an agreement with M.A.N. in Germany, to use the M.A.N. spherical-chamber direct-injection combustion system.

Briefly, in this combustion system for direct-injection diesels, the injector sprays fuel into a spherical cavity in the centre of the piston crown, covering this with a thin liquid film. The film then vaporises progressively, assisted by rapid swirling of the air charge which has been induced and accelerated through a directional port at the inlet valve, so that combustion takes place more gradually, with comparatively low rates of pressure rise. Besides smooth combustion the M.A.N. system also has the advantage of being quite tolerant of fuel quality and viscosity.

Except for the two latest V8 engines, current Berliet engines fall into two groups having cylinder bores of 120 mm (4·72 in) and 140 mm (5·51 in). In the smaller size group are in-line four-, five- and six-cylinder units, with a 140 mm (5·51 in) stroke and maximum output of 30 bhp per cylinder at 2 100 rev/min. The larger group contains normally aspirated and turbo-charged straight sizes of 160 mm (6·30 in) stroke producing 40 and 52·5 bhp per cylinder, together with a turbocharged V8 rated at 60 bhp per cylinder. Maximum governed speed in each case is 1 800 rev/min. Each group also has an 'odd' model of shorter stroke, developing the same power as one engine in that group but at the higher speeds of 2 600 and 2 200 rev/min, respectively. All the Berliet engines have replaceable wet cylinder liners.

The two latest Berliet engines are V8 designs of quite different sizes. The smaller one, the V800 has M.A.N. spherical-cavity combustion and is of only 6·9 l. (421 in^3) capacity. Its bore is 100 mm (3·94 in) and its stroke 110 mm (4·34 in). At its high speed of 3000 rev/min it develops 170 bhp. The other Berliet V8 (the V825) has a capacity of 12·76 l. (779 in^3), with Berliet direct-injection combustion and develops 300 bhp at 2 500 rev/min. The bore is 125 mm (4·92 in) and the stroke 130 mm (5·12 in), having four valves per cylinder.

BÜSSING (Büssing Automobilwerke A.G., 3321 Salzgitter-Watenstedt, Brunswick, W. Germany)

For many years the diesel engines of this West German truck and bus manufacturer were of the indirect-injection type, having turbulence antechambers in their cylinder heads. However, in 1969 the company rationalised its engine range and standardised on direct-injection.

The direct-injection series covers a power spread of 135 to 310 bhp with three basic engine sizes. They include 7·42 l. (453 in^3) vertical and horizontal units of 108 mm (4·25 in) bore and 135 mm (5·31 in) stroke; 11·58 l. (706 in^3) vertical and horizontal units of 128 mm (5·04 in) bore and 150 mm (5·91 in) stroke; and a 12·32 l. (754 in^3) underfloor engine of 132 mm (5·20 in) bore and 150 mm (5·91 in) stroke. Like the earlier units that they superseded, the direct-injection engines are four-stroke in-line six-cylinder units, fitted with wet cylinder liners. The conventional toroidal-cavity combustion chamber in the piston crown has been adopted. The fuel injectors have three hole nozzles.

In the smallest capacity range, the vertical S7D and horizontal U7D develop 155 bhp (DIN), but the horizontal U7D is also available derated to 135 bhp. Both these outputs are at the maximum governed speed of 2 400 rev/min. A similar pattern has been adopted with the 128 mm (5·04 in) bore engines, where the horizontal U11D develops either 210 or 217 bhp at 2 100 rev/min. The 12·32 l. (752 in^3) U12D is a 132 mm (5·2 in) bore derivative of the U11D, and it is rated at 240 bhp at 2 200 rev/min.

Like the majority of other underfloor engines, both British and continental, the Büssing units have their cylinders lying on the right-hand side of the crank-case; the in-line fuel injection pump is installed transversely, at the front of the cylinder block, where it is driven by bevel gears from the timing train. Special attention has been paid to the cooling of the pistons and exhaust valves in all the new Büssing engines, the valves having hollow stems filled with metallic sodium and the Alfin pistons have their undersides sprayed with oil from a jet at the bottom of the bore.

In an effort to meet the newer power requirements being demanded for lorries by German law, Büssing has introduced a turbocharged version of its biggest six-cylinder engine. This is available rated at either 280 or 310 bhp, with corresponding torques of 723 or 853 lbf ft (100 or 118 kgf m), and is known as the U12DA. It has an individual head for each cylinder, and each piston has an additional oil-control ring.

CATERPILLAR (Caterpillar Tractor Co., Peoria, Illinois, U.S.A.)

For automotive installation this American maker concentrates on four-stroke diesels of in-line six and V8 form. The latest designs attain a high standard of reliability, and altogether a power span from 150 to 375 bhp (SAE) is covered.

The in-line engines are unusual, relative to modern general world-wide practice, because they have air cell combustion chambers screwed into the cylinder head. Fuel economy suffers, but there are gains in cold-starting performance through the use of glow plugs and in simple construction of the fuel injectors using a single 0·029 in (0·73 mm) diameter hole. These injectors are termed capsules as they are about 2 in (50·8 mm) long and an inch (25·4 mm) in diameter. To withstand the hot blast from the orifice of the pre-combustion chamber, each piston has a stainless steel plug in its centre. The underside of the pistons is cooled by an oil jet from the inside of the crankcase, and wet cylinder liners are used. The pistons have two compression rings in a cast-iron carrier, the oil-control ring being spring loaded. The crankcase-mounted camshaft is driven by helical gears at the front of the engine, and the oil pump and water pump are among the units driven from the same timing gear-train. Aluminium bearings are used for the crankshaft journals, and a water-cooled heat exchanger is included in the full-flow lubrication circuit.

These 'sixes' are made in two sizes—121 mm (4·75 in) bore by 152 mm (6 in) stroke, giving 10·5 l. (641 in^3), and 137 mm (5·4 in) bore by 165 mm (6·5 in) stroke, giving 14·6 l. (893 in^3). Both are turbocharged, but both are also available with an intercooler to raise power and efficiency. The smaller engine is governed at 2 200 rev/min and the larger at 2 100 or 2 000 rev/min according to model. In both cases the speeds are comparatively fast for long-stroke engines.

In straight turbocharged form the 10·5 l. (641 in^3) Cat is known as the 1673C and is rated at 225 bhp (SAE). When intercooled its power rises to 270 bhp, and fuel consumption improves over the upper end of the speed range at the expense of economy at the lower speeds. Minimum consumption is 0·392 lb/bhp h (178 g/bhp h). The 14·6 l. engine, which, at about 3 300 lb (1 500 kg) is over 1 000 lb (454 kg) heavier than the 10·5 l. engine, develops 325 bhp (SAE) with straightforward turbocharging, or 375 bhp when intercooled. Corresponding maximum torques are 1 000 lbf ft (724 kgf m) and 1 145 lbf ft (829 kgf m) and the minimum specific fuel consumptions, 0·378 lb/bhp h (178 g/bhp h) and 0·37 lb/bhp h (168 g/bhp h)—in this case the intercooled engine consistently has a slightly better fuel economy.

The V8s are made in three sizes, 8·56 l. (522 in^3), 9·4 l. (573 in^3) and 10·4 l. (636 in^3). All have 114 mm bore (4·5 in). The first is oversquare (104 mm or 4·1 in stroke); the second is square, whilst the third is undersquare (127 mm or 5 in stroke). These moderate strokes enable the 8·56 l. to peak at the high speed of 3 200 rev/min and the 10·4 l. at either 2 800 or 3 000 rev/min according to the power setting. Though these engines, termed the 1100 series, are naturally aspirated, they are each available at alternative power settings. They cover a considerable 'bracket' between 175 bhp max (SAE) and 225 bhp max; the net ratings of the trio extend from 137 bhp to 208 bhp, according to duty requirements. These V8 Caterpillars are direct-injection designs of European style, having straight-sided toroidal-cavity combustion chambers. As a result, minimum specific fuel consumption is around 0·37 lb/bhp h (168 g/bhp h). They are compact engines, being only 36 in (914 mm) long, 34 in (867 mm) deep and 31 in (789 mm) wide, and weighing only 1 196 lb (544 kg).

Although their general layout is orthodox, the Caterpillar V8s have several notable internal features. The four-hole fuel injectors are very long and slim, with remote-seat needle valves. They are buried perpendicularly in the deep cylinder heads and protected by the rocker covers. Induction tracts are incorporated in the heads; there are no separate induction manifolds. There are no cylinder liners. The low-drag pistons have only two rings, a spring-loaded chromed oil-scraper and a molybdenum-coated compression. Both are above the gudgeon pin. The compression ring is in a cast-iron carrier and has more side clearance than usual to allow it to twist and give additional oil scraping on induction and exhaust strokes.

Bearings are of aluminium alloy—a Caterpillar tradition. The bolts securing the main bearing caps are inclined at 30 degrees to give extra transverse strength. A six-lobe oil pump, rather than a meshing-gear design, is used delivering up to 24 gal/min (109 l./min). A water-fed heat exchanger is used in the lubrication circuit.

CHRYSLER U.K. [Chrysler (United Kingdom) Ltd., Dunstable, Bedfordshire, England]

In 1954 when Tilling-Stevens Ltd. first introduced its three-cylinder two-stroke diesel engine with opposed pistons operating the crankshaft via rocker-levers, quite a stir was caused, together with some scepticism. The design has vindicated itself, however, and is acknowledged to have a long life as well as offering good fuel consumption and comparatively light weight. When the Rootes Group took over Tilling-Stevens it acquired a valuable asset in the TS3, as the engine was called. It is now termed the 3D-215, and is now in the hands of the Chrysler Corporation, which took over Rootes and renamed it Chrysler UK.

The bore has been increased since 1954, so that it is now 3·375 in (86 mm) increasing the swept volume from 3·25 to 3·52 l. (202·3 to 216·9 in^3). This has boosted the power to 135 bhp at 2400 rev/min and the torque to

The three-cylinder horizontally-opposed Chrysler two-stroke of the rocking beam type

335 lbf ft (465 kgf m); usually its power is derated to 125 bhp at 2 300 rev/min. Fuel consumption is 0·375 lb/bhp h (170 g/bhp h) and stays below 0·4 lb/bhp h (181 g/bhp h) from 800 to 2 200 rev/min. The engine weighs only 1 075 lb (487·5 kg).

The Chrysler two-stroke diesel engine employs the Kadenacy principle in which induction air under pressure flows along the cylinder from end to end and helps to drive out the exhaust gases from the previous combustion. With the opposed-piston design there are neither valves nor a gasket. Air is admitted and exhaust driven out through ports in the removable cylinder barrel, these ports being opened and closed at the appropriate time by the pistons. The pistons transmit their power to the crankshaft via double-armed rocker levers at each side. Induction air is blown in by a Roots lobe-blower mounted on the front-end of the engine. A distributor fuel injection pump is now used, which is driven through a gear train at the rear end of the crankshaft.

Industrial and marine versions of this two-stroke are made. For industrial use constant-speed ratings of 1 500 and 1 800 rev/min are available,

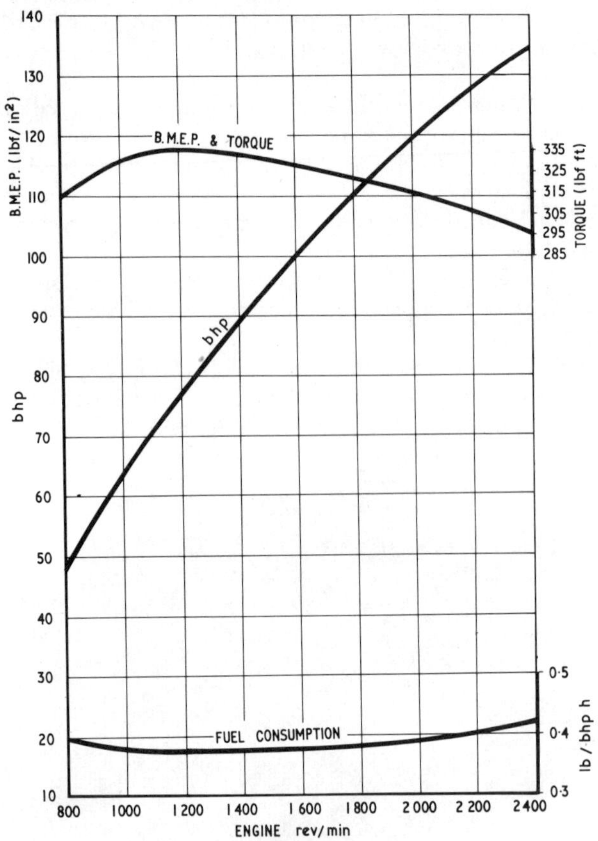

Performance curves of the 3·52 l. Chrysler diesel at its peak setting

as well as an all-speed model giving up to 2 000 rev/min. For marine work, speeds up to 2 200 rev/min are allowed.

An incidental advantage of the opposed-piston two-stroke is that it can run on a wide variety of petroleum-based fuels without fuss. When lubrication is arranged for the fuel injection pump from the engine's oil system, there is no difficulty in running the engine on petrol.

CUMMINS (Cummins Engine Co., Columbus, Indiana, USA and at Coombe House, New Malden, Surrey, England)

Automotive diesel engines from the US Cummins company date back to 1930 and were among the pioneer applications of diesel power to road transport. The Cummins engine was pioneering in an important technical respect as well—its toroidal-shape combustion chamber with direct-injection. Originally the cone of the toroidal-cavity was like a pepper-pot tip, made of heat-resistant steel. The aim was to get smoother combustion from air that was trapped in the tip, escaping through the holes. This idea was eventually discarded, and a normal piston, but with a shallow cavity by most standards, is now used.

All sizes of Cummins engines have four valves per cylinder to afford a good compromise between volumetric efficiency and gas speeds high enough to get vigorous swirl. The injectors are mounted vertically down the centre of the groups of valves and have more holes than usual—seven or eight. The whole injection system, called the PT, is exclusive to Cummins. In this, each camshaft-actuated injector is not merely a valve but acts as a pump plunger as well. Fuel is constantly circulated to all the injectors by a gear-pump driven from the engine timing gears. The faster the engine goes the more fuel is circulated and the higher the pressure. The wider the delivery orifice is opened the higher the pressure delivered to the injectors as well. The consequent fuel pressure in the injectors determines how far the needle plungers can be lifted against their springs. Then the cam on the camshaft comes round and pushes down the plunger in the injector at the appropriate time, forcing the fuel trapped under the needle-plunger through the multi-hole nozzle into the combustion chamber.

Until a few years ago, all Cummins diesels were four- and six-cylinder in-line designs. Calls from users for compactness coupled with extra power, however, caused vee-form designs to be introduced. The first V6 and V8 diesels with $5\frac{1}{2}$ in bore and $4\frac{1}{8}$ in stroke, termed the Vim and Vine engines were large ones with respective maximum power ratings of 192 and 254 bhp developed at the comparatively high speed of 2 600 rev/min. This compact, oversquare high-revving theme was then expanded to a new lightweight range of vee-form engines. Two designs have emerged; the 5·8 l. (359 in^3) Val V6 and the 7·7 l. (449·8 in^3) Vale 8. These are 30% oversquare and develop 134 and 177 bhp, respectively, at 3 300 rev/min and weigh only 1 020 lb (462 kg) and 1 220 lb (553 kg) despite being mainly of cast iron. As with all Cummins engines, wet cylinder liners are used. Slightly larger capacity versions of these lightweight vee-form engines, termed the V6-155 and V8-210 have since been produced. By increasing the stroke by $\frac{1}{4}$ in (6·35 mm) the swept volumes have been increased to 6·2 and 8·27 l. (378 and 504 in^3). The respective powers and torques on the BS basis are 155 bhp with 302 lbf ft (41·6 kgf m) and 210 bhp with 405 lbf ft (55·8 kgf m). Now a longer stroke ($4\frac{1}{8}$ in, 104·7 mm) 9·1 l. (555 in^3) is made too. These vee engines have not yet been blown, but Cummins experience in the USA is very extensive.

In Europe, the most popular Cummins in-line engines have been the NH180 and NH220 12·17 l. (743 in^3) of 173 and 212 bhp respectively, and lighter versions, with more aluminium content have been recently developed in Europe. The in-line engines are made at the Shotts factory, in Scotland, while the vee-form engines are made in Darlington, England. The in-line engines are being fitted to shunting locomotives by most of the British rail-traction manufacturers, as well as being used for industrial and marine applications.

With the advent of the British Standard method of measuring diesel horsepower the power ratings and designations of the in-line European engines have been altered and a K added to the model identifications to distinguish them. In the revision, the 173 bhp (NH180) engine has been discontinued; this being a reflection of the upward trend in power demand for automotive applications. The only 12·17 l. Cummins available from Britain is now the NHK-205, and this is the only six-cylinder Cummins with a $5\frac{1}{8}$ in (130 mm) bore. All the other Cummins in-line engines in the NHK series have a $5\frac{1}{2}$ in

(140 mm) bore. The power range of this basic 5½ in bore 14 l. (854 in^3) engine is now from 230 to 370 bhp (BS), turbocharged from 250 bhp upwards. In the NK series the last figure of each engine designation indicates the horsepower to British Standard—thus the NTK-270 is the turbocharged 14 l. rated at 270 bhp (BS). For the future a 927 in^3 (15 l.) six-cylinder engine in a 300 to 600 bhp range of models is being developed in the USA. Weight is claimed to be not much more than for the NH engines.

DAF (Van Doorne's Automobiel Fabrieken, Eindhoven, Holland)

The DAF concern began diesel-engine manufacture by making Leyland engines under licence, and the engines still bear much similarity with the British designs even though they are made in DAF's own engine factory.

The first Leyland engine to be made under licence was the 0·375 5·75 l. six-cylinder diesel and it is still in production. However, it was not long before

Tuned-length induction pipes on the high-rated DAF DK 1160 give air-ram effect above 1 600 rev/min

DAF introduced a turbocharged version of this engine, with a net power output of 150 bhp as against 110 bhp when normally aspirated.

Smaller and bigger engines than the 5·75 l. unit have since followed. A 4·77 l. six-cylinder engine was created by making a short-stroke (100 mm instead of 120 mm) version of the 5·75 l. design. Then there is a bigger-bore

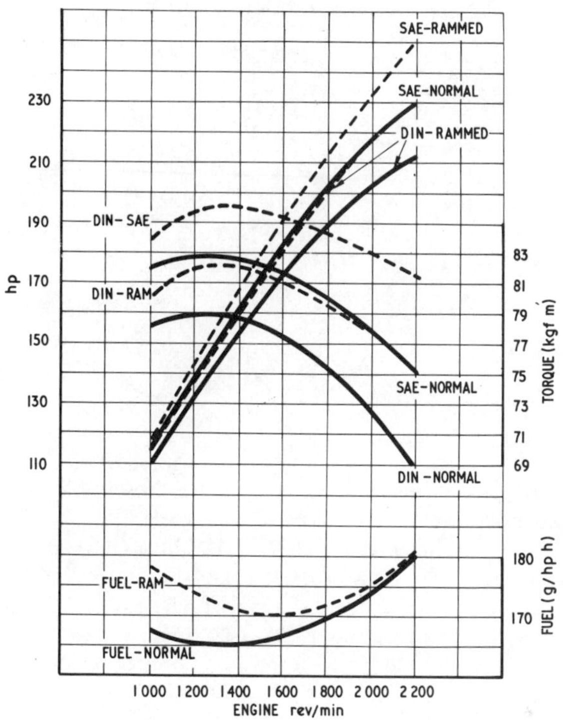

Comparative power curves between an ordinary 11·6 l. DAF and one with air-ram tuned inlet pipes

version of the 575 engine, called the 615, analogous to the Leyland 400 derived from the smaller-bore 370.

After some years of using the 11·1 l. Leyland 680 engine in its heavier trucks, DAF developed its own version in 1968 and increased the bore slightly at the same time, creating a unit of 11·6 l. capacity. This is available in 'soft' and 'tuned' forms. The normal power ratings are 165 bhp (net) at 2 000 rev/min or 212 bhp at 2 200 rev/min. In tuned form the power is raised to 230 bhp at 2 200 rev/min, although this is at the expense of fuel consumption. The tuned model has individual induction pipes sweeping over the tops of the rocker covers to a longitudinal air silencer on the cold side of the engine. A Bosch fuel pump is fitted and an oil-water heat exchanger is included in the specification. Each Alfin piston has three chromium-plated compression rings and two scraper rings. Now, to meet German power/weight legislation, the 11·6 l. DAF has been turbocharged to 304 bhp, with a corresponding high torque of 813 lbf ft (108 kgf m).

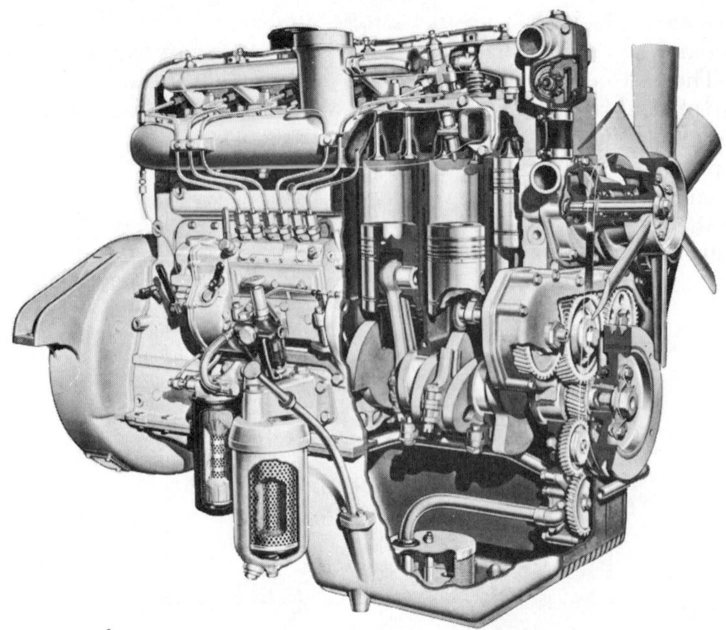

The 8·25 l. (503 in³) DAF six-cylinder engine

An 8·25 l. engine is the latest DAF six-cylinder diesel. This design has the inlet and exhaust ports on opposite sides of the heads, the injectors being outside the rocker covers. The power output is 156 bhp (net) at 2 400 rev/min. The bore is 118 mm (4·646 in) and the stroke is 126 mm (4·96 in). As with the other engines, a nitrided crankshaft, copper/lead bearings, front-end timing gears, cast-iron cylinder block and dry liners are among the features of the design. There is now a turbocharged version of this engine, also, developing 201 bhp.

DAIMLER-BENZ (Daimler-Benz A.G., 7 Stuttgart-Untertürkheim, W. Germany)

Pioneering work by Daimler-Benz in the automotive diesel field had its influence on British as well as German road transport, for one of the first diesel buses in Britain was a Karrier six-wheeler with a 77 bhp 7 l. (427 in³) Daimler-Benz engine used by Sheffield Corporation. The development of Daimler-Benz engines centred around the antechamber layout of indirect-injection until 1967, when, except for the small car engines, a switch was made to direct-injection, using bath-tub piston cavities.

None of the in-line engines has cylinder liners. Pistons have three wedge-section compression rings (the top one chromium-plated) and one slotted scraper ring—all above the gudgeon pin. Oil passes through a single heat exchanger consisting of a coiled pipe within a cylindrical water jacket. Generous water space is provided around the cylinders and there are valve-seat inserts in the cylinder heads. Helical gears at the front provide the timing drive, while crankcase bottom flanges are on the crankshaft centre line.

There are six engines in the commercial vehicle range. The two smallest, OM314 and OM352, have a bore of 97 mm (3·82 in) and a stroke of 128 mm (5 in). The OM314 is a 3·78 l. 'four' (231 in^3) and the OM352 a 'six' of 5·6 l. (346 in^3) capacity, with respective powers of 85 and 130 bhp (140 SAE) at 2 800 rev/min. There is also a supercharged version of the six-cylinder engine which produces 156 bhp at 2 800 rev/min.

The same 128 mm (5 in) stroke had been used for a larger-bore six-cylinder engine the OM327, of 7·98 l. (486 in^3) and a power of 170 bhp. However, in 1969 this was replaced by the completely different OM360. This

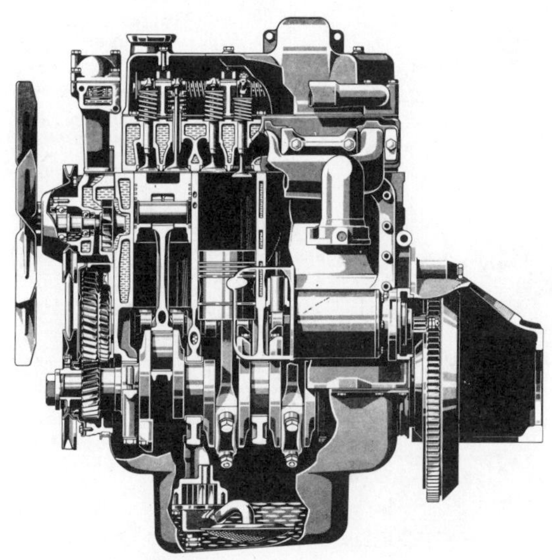

Section through the OM 314 four-cylinder engine

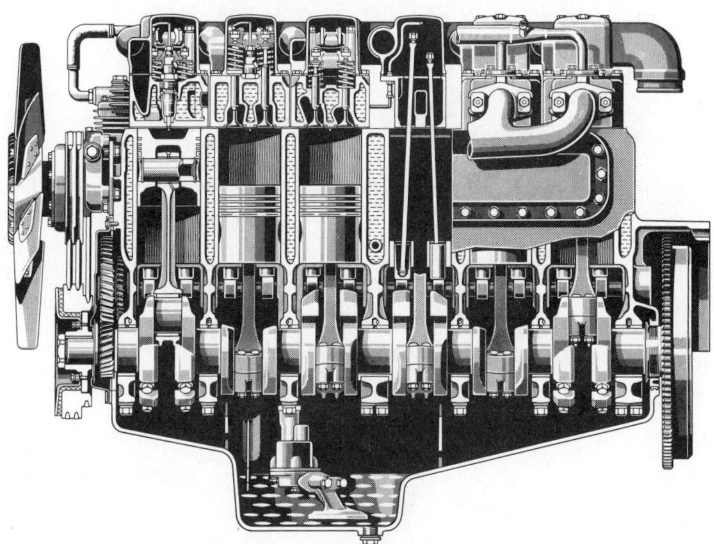

Section through the OM 355 240 bhp (255 bhp SAE) Daimler-Benz six-cylinder engine

8·12 l. (495 in^3) engine has a 115 mm (4·53 in) bore and 140 mm (5·51 in) stroke. It is slower revving, governed to 2 200 rev/min; its power setting is 170 bhp, compared with the 192 bhp when it is governed at 2 500 rev/min. At either rating the peak torque is 59 kgf m (427 lbf ft).

The two big engines in the range, the OM346 and OM355, share the same bore of 128 mm (5·03 in). but the strokes differ. The OM346 has a 140 mm (5·5 in) stroke, whereas the OM355 has a stroke of 150 mm (5·9 in). Two power settings are available for the smaller engine—185 or 210 bhp

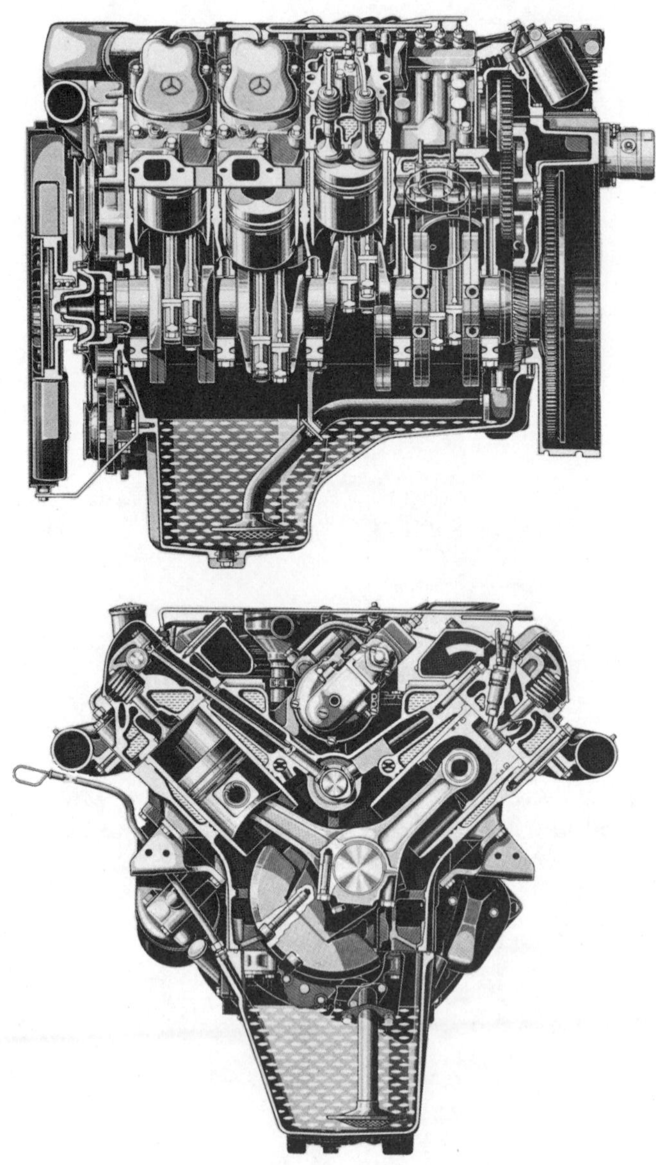

Longitudinal (top) and transverse (bottom) sections of the Daimler-Benz OM 402 256 bhp V8

(DIN) at 2 200 rev/min. Lessons learned from getting the extra power from the 346 engine have been put to good use in the long-stroke version. Each cylinder has its own individual head with four valves operated by forked rockers. A viscous fan-drive coupling with thermostatic control is a standard fitting. The larger engine of this specification produces 240 bhp (or 255 bhp SAE) at 2 200 rev/min. The minimum specific fuel consumption of 0·348 lb/bhp h (157·8 g/bhp h) is good.

Considerable success has been scored by Daimler-Benz in small diesel engines for cars and taxis. For many years the standard unit was a 1·8 l. (107 in^3) four-cylinder engine for these duties. But now a 2·2 l. (134 in^3) diesel is used. This is oversquare (bore 111 mm (4·42 in), stroke 92 mm (3·65 in)). This allows a relatively high governed speed—4 200 rev/min at which 60 bhp (64 bhp SAE) is developed. It has an in-line injection pump. Unlike

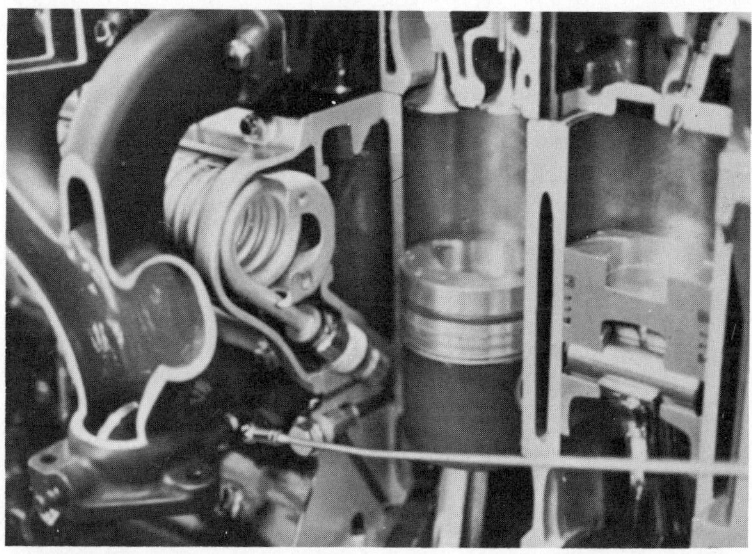

Section through part of one of the direct-injection Daimler-Benz engines showing the coiled oil–water heat exchanger and bath-tub cavity pistons

other Daimler-Benz engines, this model, called the OM615, has dry cylinder liners, otherwise it follows car engine practice. The timing drive is by chain and it weighs 405 lb (183 kg).

For heavy duty transport and off-road earthmoving work, as well as marine and industrial duties, a range of 90 degree vee-form engines, in six-, eight- and ten-cylinder forms is now offered by Daimler-Benz. They are known as the MB833, 837 and 838. The height and width are the same for every model, just the length varying between 43 and 61 in (1 092 to 1 549 mm), according to the number of cylinders. Mechanically supercharged or turbocharged, as well as the normally aspirated models, are available, and the power range afforded is from 260 bhp at 1 500 rev/min to 750 bhp at 2 200 rev/min.

The latest Daimlar-Benz engines are a V10 and a V8; a V12 model in the family is planned. The V10, the OM403, is a 90 degree vee-form of 15·95 l. (973 in^3) capacity, having a bore of 125 mm (4·92 in) and a stroke of 130 mm (5·12 in). Using 10 cylinders has kept the piston speed acceptable and yet avoided going oversquare in order to achieve a governed speed of 2 500 rev/min, at which the DIN bhp is 320. The maximum torque is 103 kgf m (745 lbf ft) at 1 600 rev/min. A good minimum specific fuel consumption is claimed —147 g/bhp h (0·324 lb/bhp h). Simple bath-tub piston cavities are used. There is an individual head for each cylinder, the timing drive is by gears at the rear and, unlike the in-line engines, these vees have wet cylinder liners. The V10 engine weighs less than a ton. The V8 develops 256 bhp.

DEUTZ (Klöckner-Humboldt-Deutz A.G. Werk Ulm, 79 Ulm/Donau, Postfach 543, W. Germany)

The Deutz company is probably best known in automotive circles for its air-cooled designs, but it does have a substantial range of water-cooled engines as well, mainly in the large horsepower category. All the air-cooled engines are normally aspirated, whereas all except the smallest model in the water-cooled range are turbocharged. Indirect-injection used to be the order of the day on Deutz engines and this combustion arrangement still applies to the water-cooled units. However, in the last major revision of design of the air-cooled engines, in which the bore/stroke dimensions were made more nearly square, direct-injection was adopted. A big improvement in fuel consumption resulted, so that all the Deutz air-cooled engines now hit a minimum specific fuel consumption of 160 gm (0·362 lb)/bhp h.

Production is now being concentrated on two basic ranges of air-cooled diesels for automotive applications. One is an in-line range with four or six cylinders of 100 mm (3·94 in) bore and 120 mm (4·73 in) stroke. These have

An air-cooled four-cylinder Deutz

capacities of 3·768 and 5·653 l. (280 and 345 in^3), with DIN of 70 or 80 bhp and 110 or 120 bhp at 2 800 rev/min. The other range consists of 90 degree vee-form engines in six-, eight-, ten- and twelve-cylinder forms with a common bore and stroke of 120 mm by 125 mm (4·73 by 4·92 in). Capacities range from 8·4 to 16·9 l. (517 to 1 035 in^3), with respective power outputs of 170, 232, 290 and 340 bhp at 2 650 rev/min. Lower power settings are used for buses. For their capacities, these vee-form Deutz air-cooled engines are not heavy, even the V12 weighs barely a ton (1 015 kg).

Cast-iron cylinder blocks and individual aluminium heads with no gasket between are the pattern of Deutz air-cooled engine design. Induction-hardened crankshafts with copper/lead bearings are used and there are two

A Deutz turbocharged and intercooled V12 unit; note the use of two turbochargers and the cooler above the vee between the two cylinder banks

valves per cylinder. Timing drive is by gears in all cases, but whereas the in-line engines have the timing drive at the front-end, the vee-form engines have it at the rear. There are injector-nozzle differences between the two types as well; the in-line engines have four-hole nozzles while the vee-form engines have two-hole nozzles. The cooling arrangement consists of a large shrouded axial blower with fixed intake-guide vanes in front of it. The finned cylinders are completely cowled to get the maximum effect from the cooling air.

A bore of 135 mm (5·32 in) and a stroke of 160 mm (6·3 in) apply to all the Deutz water-cooled engines in the F716 automotive series. All have wet cylinder liners. The four-, six- and eight-cylinder models are in-line and have respective capacities of 9·16 l., 13·74 l. and 18·32 l. (559, 838 and 1 118 in^3). The 12- and 16-cylinder models are of 120 degree vee formation and are so big, having capacities of 27·48 l. and 36·64 l. (1 680 and 2 233 in^3), that they are suitable in automotive form only for specialised heavy transport and off-highway vehicles.

DORMAN (Dorman Diesels Ltd., Tixall Road, Stafford) (part of the General Electric Co. Group)

Although Dorman engines were common fitments for commercial vehicles in the early 1930s (the first diesel having been designed in 1929) since 1938 the company has concentrated on marine and industrial designs. There is now a wide range of Dorman engines for industrial use, with powers from 6 to 930 bhp and cylinder combinations from 1 to 12, air-cooled as well as water-cooled.

In 1970, a period of experimental installation of six-cylinder engines in trucks culminated in the introduction of a 13·28 l. (810 in^3) V8 direct-injection diesel designed with automotive applications primarily in mind. Its gross output is claimed to be 235 bhp at 2 200 rev/min and the peak torque to be 640 lbf ft (88·5 kgf m) at 1 550 rev/min. It weighs 2 950 lb (1 340 kg), which is rather heavy for a normally aspirated automotive engine of this size, but turbocharged models are planned. Indeed, the 8JV, as the Dorman V8 is called, is said to be just the first of a multi-purpose series of engines with various numbers of cylinders developing up to 600 bhp.

To keep piston speeds within acceptable limits an oversquare bore/stroke ratio has been used; the bore is 130 mm (5·12 in) and the stroke is 125 mm (4·92 in). The pistons have toroidal combustion cavities. They have two compression rings and one oil-control ring. Wet cylinder liners are used, sealed by three O-rings at the bottom. There is a separate head for each cylinder, and two valves per head; while the timing drive is by a train of helical gears at the front.

The connecting rods are paired on the crankshaft journals and so that the connecting rods can be withdrawn through the cylinder bores they are split diagonally. Main bearings are 115 mm (4·53 in) and big end bearings are 95 mm (3·74 in) diameter. Bearings are aluminium/tin. The engine's b.m.e.p. is 120 lbf/in^2 (8·44 kgf/cm^2) and its minimum specific fuel consumption 0·362 lb/bhp h (160 gm/bhp h).

FBW (A.G. Franz Brozincevic & Co., Wetzikon, Zurich)

Pioneer of the horizontal diesel engine in Switzerland, and still employing units of this type to propel goods as well as passenger vehicles, this company makes two basic power units of such type, both having their six cylinders arranged in line on the left of the crankcase. This type of layout was later adopted by other Swiss manufacturers of horizontal engines. An unusual feature of the larger engine, which is also available in vertical form, is its use of two masked inlet valves along with one exhaust valve per cylinder.

All FBW engines have replaceable wet cylinder liners and hardened crankshafts, while their timing gears are driven by a chain which has an automatic tensioning device. Again, all are of four-stroke direct-injection type, utilising cavity pistons, compression ratios varying from 17·3 : 1 to 17·8 : 1.

Goods models of 11 tons gross weight (11·1 tonnes) are powered by the 6·99 l. (426 in^3) type CU underfloor engine, having a 106 mm (4·17 in) bore and 132 mm (5·19 in) stroke, which develops 130 bhp at 2 400 rev/min. Heavier goods vehicles of 13 and 15 tons (13·2 and 15·2 tonnes) gross have this same engine, but in these cases in turbocharged form. Then, as type CUA, its power output is lifted to 160 bhp. Corresponding figures for maximum torque (at 1 600 rev/min) are 44 kgf m (318 lbf ft) and 57 kgf m (412 lbf ft).

Further up the scale comes type E, which is of vertical configuration, and its underfloor equivalent, type EU. These have bore and stroke dimensions of 125 mm (4·92 in) and 150 mm (5·90 in), thereby increasing the capacity of this six-cylinder design to 11·045 l. (673 in^3), and, when normally aspirated, the maximum output common to both is 172 bhp at 1 900 rev/min and with a torque of 524 lbf ft (72·5 kgf m) at 1 200 rev/min. This engine is now available turbocharged, which raises the power to 230 bhp.

Subsequently the type E engine has been bored out to 128 mm (5·04 in), increasing its capacity to 11·58 l. (707 in^3). At the same time, for yet more power, maximum speed has been raised to 2 100 rev/min at which the output is 210 bhp. Maximum torque of 600 lbf ft (82·9 kgf m) is produced at 1 400 rev/min. There is also an underfloor version of this engine, designated type EU3. Fuel consumption is 162 g/bhp h (0·356 lb/bhp h) but it increases fairly steeply at the higher speeds, reaching 182 g/bhp h (0·4 lb/bhp h) at 2 100 rev/min.

FIAT (Fiat S.P.A., Turin, Italy)

Toroidal-cavity combustion chambers in the piston crowns are featured in all engines of Fiat make, which, since 1960, have been made to operate on the direct-injection principle. The combustion chamber is offset slightly to the right of the piston's vertical axis and served by an inclined injector which can be removed without taking off the rocker cover protecting the valve gear. The injector nozzle has four holes.

At the 1968 International Motor Show, at Geneva, Fiat introduced a new range of rationalised lightweight diesels for goods chassis rated from 3½ to 9½ tons gross, that is in metric terms from 3·5 to 9·6 tonnes. Known as the 800 Series, it comprises three in-line vertical units of normally aspirated form with three, four and six cylinders, each cylinder being fitted with a dry liner having a bore of 95 mm (3·74 in) and stroke of 110 mm (4·33 in) to give a swept volume of 779·77 cm^3 (47·5 in^3) per cylinder. The engines are governed to a top speed of 3 200 rev/min at which they deliver 51, 70 and 110 bhp. Fuel injection pumps are of in-line camshaft type. The engines, which have many parts in common, are also for use on agricultural tractors.

Until the end of the 1960s the remaining Fiat engines, all straight sixes, fell into three groups having respective bore and stroke dimensions of 120 mm (4·72 in) and 135 mm (5·31 in), 130 mm (5·12 in) and 145 mm (5·71 in); and 135 mm (5·31 in) and 150 mm (5·91 in), providing total displacements of 9·161 l., 11·548 l. and 12·883 l. (559, 705 and 785 in^3). The first group was governed at 2 400 rev/min to produce 168 bhp (vertical) and 156 bhp (horizontal), while the larger engines were restricted to 1 900 rev/min, at which the vertical versions developed 177 and 208 bhp and the underfloor versions 173 and 198 bhp. Design features of the engines in these three groups included dry cylinder liners, copper/lead bearings and induction-hardened crankshafts—and these are still Fiat features.

These engines were supplanted by more powerful developments in 1969.

The power of the 11·548 l. (705 in^3) engine was raised to 194 bhp, and a six-cylinder 9·82 l. (598 in^3) design was introduced; this has a bore of 122 mm (4·8 in) and a stroke of 140 mm (5·51 in). It develops 200 bhp at 2 500 rev/min and a maximum torque of 70 kgf m (506 lbf ft) at 1 200 rev/min.

In 1970, the demand for a more powerful engine for the top-weight lorries was met by the introduction of a new six-cylinder design, rated at 260 bhp (DIN), which is governed at 2 200 rev/min. Its bore is 137 mm (5·39 in) and its stroke 156 mm (5·69 in), giving a cubic capacity of 13·8 l. (844 in^3). It develops its maximum torque of 101 kgf m (732 lbf ft) at the unusually low speed of 900 rev/min.

The timing drive is at the front and a water-cooled heat exchanger is bolted directly to the twin filters of the lubricating-oil circuit. The fuel pump is on the exhaust side of the engine and the exhaust manifold is therefore shielded, with the injection pipes routed over the top of the heat shield.

FODEN (Fodens Ltd., Elworth, Sandbach, Cheshire, England)

One of the most interesting engines produced by the British transport-vehicle industry is the Foden two-stroke diesel, produced in four- and six-cylinder forms. It is notable for delivering a great amount of power from a comparatively small capacity (which has been increased over the years).

The in-line six-cylinder engine is the most popular in the range and has found even wider application as a marine than as an automotive engine. Governed at 2 250 rev/min, the 4·8 l. (293 in^3) Foden two-stroke delivers 180 bhp net. When first introduced in 1948 its capacity was 4·1 l. (250 in^3) and it developed 126 bhp at 2 000 rev/min. The engine weighs only 1 270 lb (576 kg) without electrical equipment.

In general configuration the Foden two-stroke is conventional except in details such as individual tuftrided cast-iron heads and timing gears at the flywheel end of the crankshaft. Wet cylinder liners are used in an aluminium-alloy block. Air is blown by a Roots lobe compressor through tangential ports around the bottom of the cylinders and exhaust gases flow through pairs of push-rod operated poppet valves at the top of the cylinders. A normal in-line fuel injection pump is used, and this is positively lubricated from the engine's oil system. The long-skirted pistons have toroidal cavities and are of cast iron. A fire ring at the top supplements the normal piston rings.

For marine installations a special double-banked engine unit has been developed, consisting of two Foden six-cylinder engines geared together, side by side, driving a common reduction gearbox.

Fodens have obtained even greater power-to-weight ratios by turbocharging the two-stroke diesels. Supercharging in this way raises the maximum power of the 4·8 l. engine to 225 bhp. The pattern of turbocharging is more advanced than most because it embodies intercooling in which the charge air is cooled and therefore densified by being passed through an air-to-air radiator before entering the cylinders.

The weight of this engine is still less than 1 500 lb (680 kg) complete, and its minimum specific fuel consumption improves from 0·395 to 0·37 lb/bhp h (179 to 167 g/bhp h). The specific power figures of 47 bhp/l. and 6·3 lb/bhp are probably the best combination of any automotive diesel in the world at present.

The 225 bhp turbocharged version of the 4·8 l. Foden two-stroke six-cylinder engine

FORD (Ford Motor Co. Ltd., Dagenham, Essex, England)

Agriculture was Ford's first market for its own diesel engine. This was a 3·6 l. four-cylinder engine for the Fordson Major tractor. It was not until March, 1954 that the company entered the strictly automotive field with a direct-injection engine based on the agricultural design. It was an immediate success and subsequent Ford diesels have been based on its design.

The original Ford diesel had wet cylinder liners. Later there was a change to dry liners and now most of the engines have no liners at all. The timing drive is at the front by helical gears and it has induction hardened crankshafts with aluminium-tin bearings. In-line fuel-injection pumps are used on all models while the four-hole injectors spray fuel into bath-tub piston cavities.

The same stroke of 4·52 in (115 mm) is used for several Ford diesel units. There are two normally aspirated six-cylinder models, which in standard form are canted to fit below the floor of the cab of the trucks, which forms the principal installation. One of these engines has the same 4·22 in (107 mm) bore as the four-cylinder engine. This is the 6·23 l. (380 in^3) model, developing 114 bhp (DIN) at 2 800 rev/min. The other six-cylinder engine is the 5·95 l. (363 in^3) with a 4·125 in (105 mm) bore. The power of this engine is 103 bhp.

The 365 with its bores machined directly in the cast-iron block, supplanted the 360, which has dry cylinder liners, but the 360 is still made in turbocharged form. This has a DIN power rating of 143 bhp at 2 400 rev/min. The maximum torque has gone up from 236 to 338 lbf ft (46·7 kgf m) compared with the naturally aspirated performance. The latest Ford diesel is an oversquare 2·36 l. (144 in^3) four-cylinder indirect-injection design offered at either 54 or 61 bhp at 3 600 rev/min. There is also an 87 bhp six-cylinder model for industrial and marine applications. They are probably the first diesels to have cast crankshafts and external toothed belt drive for the camshaft.

The Ford 2·3 l. (144 in^3) inclined four-cylinder diesel of 54 bhp or 61 bhp at 3 600 rev/min, with toothed belt drive for the camshaft

GMC (General Motors Corporation, Detroit, Michigan, USA)

On entering the diesel engine field back in the late 1930s, GMC, the world's biggest motor manufacturer, pinned its faith on the uniflow-scavenged two-stroke design. In this, the piston uncovers a row of tangential ports in the cylinder wall as it descends during each power stroke, thereby admitting compressed air which swirls around the cylinder and blows out the products of combustion through the overhead exhaust valves. A Roots blower, driven by gearing from the timing train at the flywheel end of the engine, supplies air at a pressure of about 7 lbf/in^2 (0·49 kgf/cm^2) to the air chest.

To suit this design of engine and to maintain quality control over all components, GMC also developed its own system of fuel injection, based on the use of a constant-stroke, variable-spill pump incorporated in each injector, which is operated by push-rod and rocker from the camshaft. A vane-type pump circulates fuel at 20 lbf/in^2 to the injectors, each of which meters, times, pressurises and atomises the fuel in one operation.

The first range of GMC two-stroke diesels comprised in-line units of three, four and six cylinders, each cylinder having a dry liner with bore and stroke

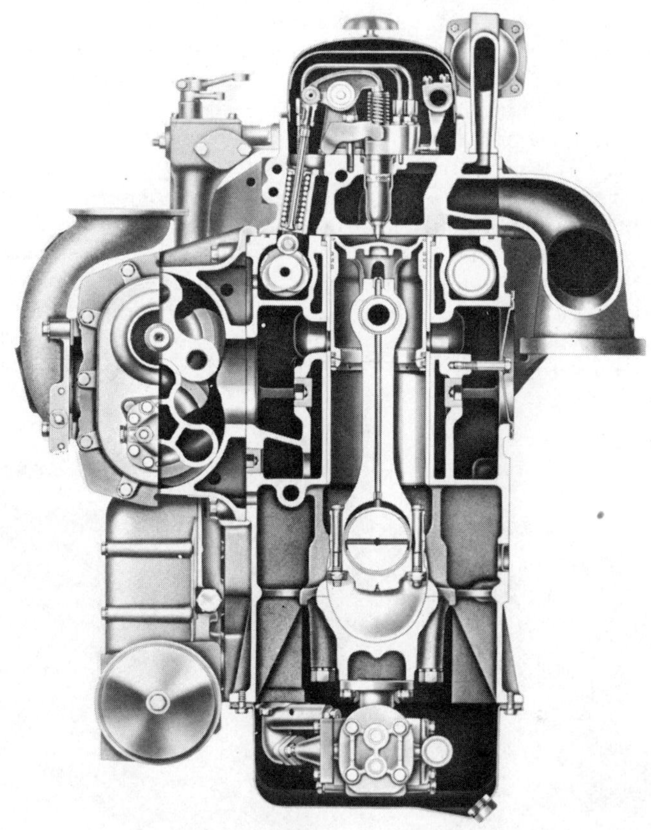

Section through an in-line GMC two-stroke diesel

dimensions of 4¼ in (108 mm) and 5 in (127 mm), respectively, to give it a swept volume of 71 in^3 (1 163 cm^3). Hence the range became known as the 71 Series. With the formation, after the Second World War, of the Detroit Diesel Engine Division, GMC engines of this type took the general name Detroit Diesel. Later on, the 71 Series was expanded to include V6, V8 and V12 engines, these again having a maximum rated output of 39·8 bhp per cylinder at 2 100 rev/min. Thus the 71 Series engines now cover maximum ratings from 106 to 475 bhp, with a power overlap by the straight-six and V6 units. The engines are also available with a high economy rating (for lowest fuel consumption and longest life) and a balanced match rating (where performance and fuel economy considerations are equally important). Cylinder banks of the vee-form engines are inclined at an angle of 63·5 degrees.

After the 71 Series came the 53 Series, comprising three- and four-cylinder in-line and V6 and V8 units of the same basic design but with bore and stroke dimensions of $3\frac{7}{8}$ in (98·5 mm) and $4\frac{1}{2}$ in (114 mm) to give a swept volume of 53 in^3 (868·5 cm^3) per cylinder. The two smallest models are governed to a maximum speed of 2 800 rev/min, at which each cylinder has a maximum power rating of 33·7 and 35 bhp, respectively, but the V6 is governed to 2 600 rev/min (32·5 bhp/cylinder) and the V8 to 2 500 rev/min (30·9 bhp/cylinder).

A recent option for Detroit Diesel automotive engines is a device called Power Control. This limits fuel input in the higher rev/min range, and thus reduces smoke and promotes greater fuel economy without impairing acceleration in the intermediate gears.

Care has been taken throughout all stages in the design of these two-stroke engines to ensure that all newly designed components are interchangeable with those of earlier versions of the same model, while up to 70% of the moving parts within a series are claimed to be interchangeable. The same attention has been devoted to ease of maintenance and servicing, for which end access ports are provided to allow quick checks to be made on the condition of cylinder liners, pistons and rings. A further feature of interest is that the symmetrical construction of the in-line engines permits the blower and other auxiliaries to be mounted on either side; and, if necessary, the cylinder block or head can

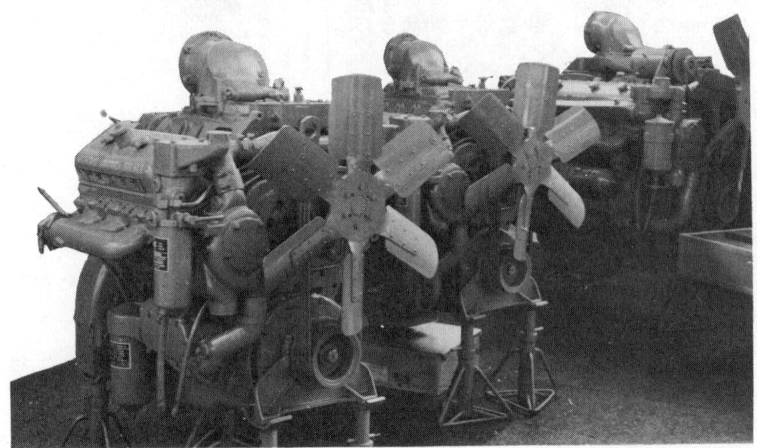

Three GMC vee two-stroke diesels of (left to right) six-, eight- and twelve-cylinder forms

be reversed. Furthermore, a simple change in the gearing can result in either clockwise or anti-clockwise rotation.

To meet competition from foreign (mainly British) diesel engines, which were then being imported in increasingly large numbers to convert existing petrol-engined medium-weight trucks to diesel power, the Truck and Coach Division of General Motors at Pontiac, Michigan, introduced the Toro-Flow V6 four-stroke diesel in 1962. This has its two banks of over-square unlinered cylinders inclined at 60 degrees to each other and is of generally conventional design. Each piston has an offset toroidal cavity in its crown, along with a recess to provide adequate clearance for each valve head, and the fuel injection system is of conventional high-pressure type, served by an American Bosch distributor pump.

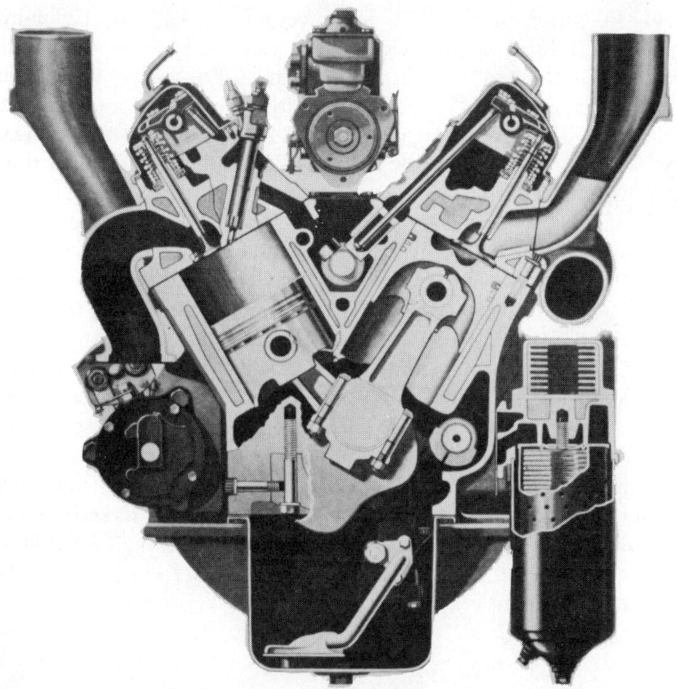

Section through the GMC Toro-Flow 60° V6 four-stroke diesel

The first engine of this type was of 748 in^3 (12·27 l.) displacement, and it was joined in 1964 by another of 351 in^3 (5·75 l.) for lighter trucks. A good feature of the design is the cylinder head, which employs angled injectors (with multi-hole nozzles) on the insides of the vee. Thus the high-pressure fuel pipes need only be short and do not have to be removed before taking off the rocker covers for tappet adjustments. In addition, individual injectors may be removed without first taking off the rocker covers.

GARDNER (L. Gardner & Sons Ltd., Patricroft, Eccles, Manchester, England)

The first British diesel engine on the heavy-vehicle market was basically a marine engine—the Gardner L2. It formed the nucleus of many conversions and its marked fuel economy, ease of cold-starting, strong pulling power and reliability were attributes which played a major part in the rapid spread of enthusiasm for diesel-engined transport in Britain in the late 1920s and early 1930s.

Demand for a special automotive design of Gardner engine soon inspired the introduction of the lighter LW series of engines in 1931. These engines, in 4·25 in (108 mm) bore, four-, five- and six-cylinder forms, are still in production, although their power outputs are now higher than when they were first introduced. The 8·4 l. (510 in^3) 6LW, for example, now delivers 120 bhp at 1 700 rev/min against 102 bhp in its early years of production.

To meet the requirements of heavier vehicles a 4·75 in (120 mm) bore, 10·45 l. (638 in^3) 150 bhp six-cylinder engine, the 6LX, was introduced in 1958. The governed speed was still 1 700 rev/min. For a more powerful version of this engine introduced in 1966, however, the governed speed was increased to 1 850 rev/min. Maximum torque of the 10·45 l. engine in this form, known as the 6LXB, went up from 485 to 536 lbf ft (74·2 kgf m) and the power to 180 bhp. No weight was added; as the Gardner engines are by no means heavy in the first place the excellent specific weight (for a four-stroke diesel) of 9 lb/bhp was attained. Even this performance has been excelled by the recently introduced 8LXB in-line eight-cylinder engine. Considering its in-line layout this engine is also relatively compact. Its power is an easy 240 bhp.

At the same time the Gardner pattern of specific fuel consumption improvement with each succeeding model was maintained. The Gardner direct-injection 6 in-stroke (152 mm) designs have always set the criterion for diesel efficiency and quality; the 8LXB raised the Gardner standards to a fresh pinnacle of 0·325 lb/bhp h (147 g/bhp h). Even at governed speed the specific fuel consumption is still a mere 0·333 lb/bhp h (150 g/bhp h). Exhaust smoke density is in the 5 to 10 Hartridge range—invisible to the naked eye.

There are many exclusive features common to all the Gardner engines. They have cast-iron blocks and heads (lapped together) with a thin steel gasket between, but the crankcase is always a liberally ribbed and flanged

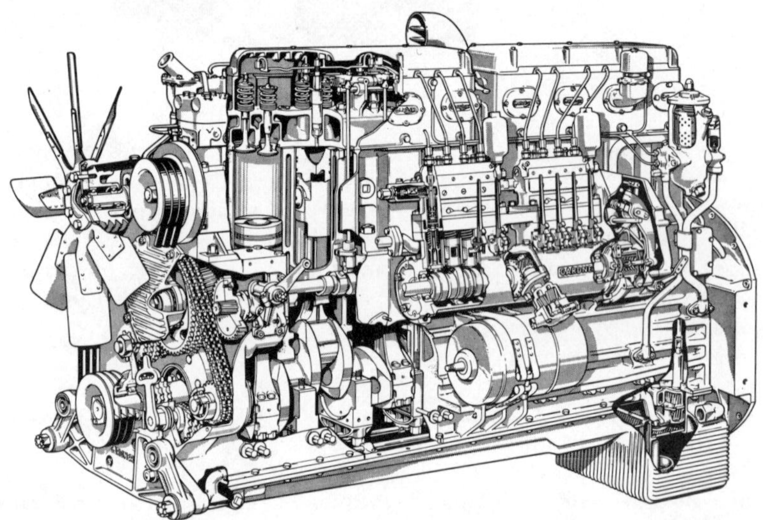

One of the world's most thermally efficient automotive diesels—the 240 bhp 8LXB Gardner straight eight

aluminium-alloy satin-finish casting. This has transverse tie bolts through the main bearing caps as well as the vertical securing studs, which extend upwards right through the block to hold the heads at the top. The whole structure is therefore put in compressive stress in all the regions of peak stress.

Crankshafts are always ground all over and are always supported by an extra roller bearing at the front, so that six-cylinder engines, for example, have eight main bearings altogether (eight-cylinder models have ten). Timing drive is by chain at the front. The inlet valves carry masks above their heads to direct incoming air into an orderly swirl within the cylinder.

Hemispherical combustion chambers are used. Four-hole injectors of Gardner design and make are served by CAV fuel injection elements actuated by a Gardner cambox. The injection pump drive incorporates an automatic advance/retard mechanism, responsive to engine load as well as to engine speed. A Gardner design of mechanical governor is fitted in which the governor-spring pressure to be opposed by the flyweights is varied according to the load-demand on the engine. This governor gives an unusually precise cut-off.

Pistons are cast from a low-expansion aluminium alloy in Gardner's own foundry. They are notable for having a very low drag factor. The skirts are unusually long and are relieved on the sides so that long stabilising areas are presented just on the thrust sides. Only two compression and one scraper-ring are fitted, all above the gudgeon pin. Deep finning is incorporated beneath the piston crowns to dissipate heat, and the undersides of the pistons are cooled by oil jets from oilways up the connecting rods. So that oil in circulation can have useful cooling value, the high-power Gardner engines are recommended to be connected to a Gardner air-to-oil radiator for mounting in front of the vehicle's cooling system radiator. All models except the 8LXB are available in horizontal as well as vertical forms.

Industrial, marine and railway-traction versions of the Gardner automotive engines are in great demand. Bigger engines designed primarily rather than incidentally, for industrial and rail-traction use are made. These are the 6L3B and 8L3B in-line six- and eight-cylinder engines of 18 l. and 24 l. (1 098 and 1 464 in^3) capacity. Their design follows Gardner tradition; the bore and stroke being 5·5 in and 5·75 in (139 and 146 mm), the governed speed is kept down to 1 300 rev/min, at which the 6L3B delivers 195 bhp and the 8L3B, 260 bhp. The corresponding torques are 823 and 1 097 lbf ft (113 and 153 kgf m).

HANOMAG (Rheinstahl Hanomag A.G., Hanomagstrasse 8, 3 Hannover-Linden, W. Germany)

Together with Henschel, Hanomag is now owned by Daimler-Benz, and so a complete reorganisation of its manufacturing policies has been carried out, including much useful rationalisation of engines and vehicles. An important result of this revitalisation has been the development of the D100 rationalised series of diesel engines in the 43 to 115 bhp (net) range.

The characteristic of all the engines in the Hanomag rationalised range is their comparatively shallow depth, which suits the forward-control vehicle installations by keeping down the engine intrusion in the cab. There are two ranges of diesel engines, but they differ only in bore; the stroke is the same (100 mm) in all cases. All engines are indirect-injection and have wet cylinder liners.

The D131, D141 and D161 (the middle figure of the model number denotes the number of cylinders) have a 95 mm (3·74 in) bore, giving respective total cylinder capacities of 2·13 l. (130 in^3), 2·84 l. (173 in^3) and 4·26 l. (260 in^3), and respective DIN power outputs of 43, 65 and 100 bhp. The D132, D142 and D162 engine series are rather more powerful, having a 100 mm (3·94 in) bore; the respective capacities of these engines are 2·36 l. (144 in^3), 3·11 l. (190 in^3) and 4·71 l. (288 in^3), with corresponding powers (DIN) of 49, 80 and 115 bhp. Governed speeds are high—four-and six-cylinder engines peak at 3 400 rev/min. The three-cylinder engines are limited to 3 000 rev/min, but this is still a high speed for engines of this type. The 131 and 132 engines are being abandoned in favour of the Daimler-Benz OM615 engine.

All the Hanomag engines have Bosch distributor pumps, driven horizontally by the timing gears at the front of the engine. Pintle-nozzle injectors, which give a conical spray, atomise fuel into the air cells. Heater plugs are fitted to facilitate cold-starting.

Copper-lead bearings support the induction-hardened crankshaft. Big ends are split at 45 degrees so that they can be withdrawn up the cylinder bores. The flat topped pistons have deep top lands, and just two compression and one oil-scraper ring are fitted above the gudgeon pin. Air is induced through passages in the valve-rocker cover and then through inlet ports which are substantially vertical in sweeping down to the valve throats.

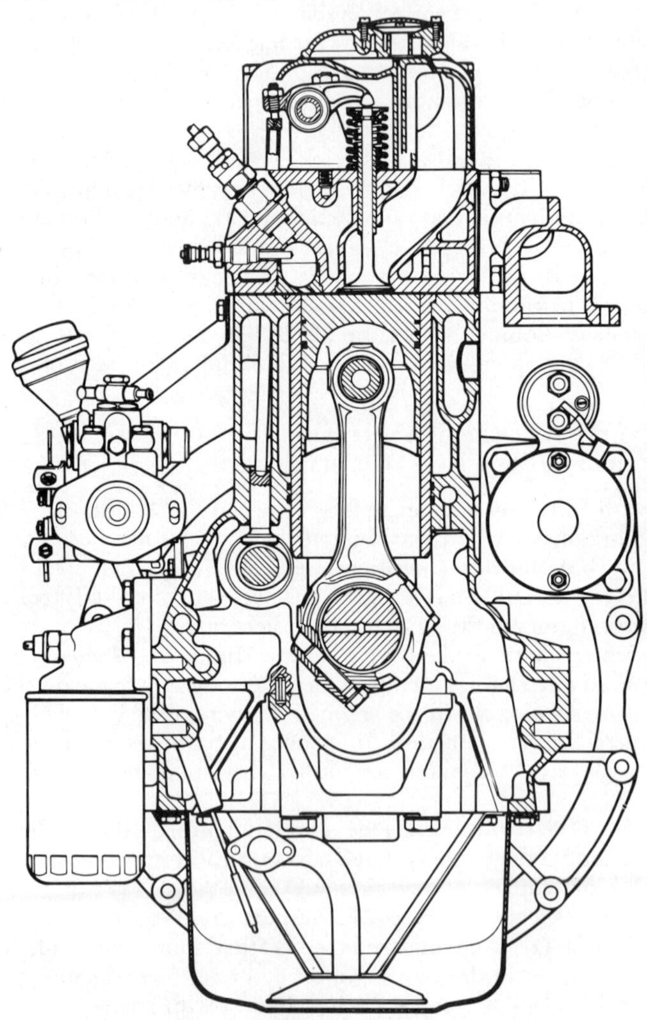

Cross-section through an indirect-injection Hanomag engine

HENSCHEL (Rheinstal Henschel A. G., 35 Kassel 2, Postfach 786)

The Henschel range of automotive diesel engines has been rationalised to two six-cylinder models of in-line configuration. Both of them are naturally aspirated, one being a 230 bhp 11·94 l. and the other a 160 bhp 7·8 l. engine. The inlet and exhaust manifolds are on opposite sides of the heads. With the bigger engine the injectors are recessed under the corners of the rocker covers so that a central location is obtained relative to the bore and yet the injectors are kept outside the rocker covers for the sake of easy maintenance.

The Bosch in-line fuel pump is driven through a rubber-buffered, flexible coupling incorporating a heavy flywheel to damp out torsional oscillations. A positive oil feed is arranged from the engine to the cam box of the fuel pump, while much use is made of flexible external pipes in the lubrication system. The fan drive incorporates a thermostatic device.

The smaller in-line six-cylinder engine has heads which cover three bores each and the injectors are to one side, still outside the rocker covers. As with the larger engine the timing drive is by gears at the front. The 7·8 l. (475 in^3)

The 11·94 l. Henschel engine installed in one of the maker's tilt-cab trucks

engine has a 115 mm (4·52 in) bore and 125 mm (4·93 in) stroke. It develops a maximum torque of 57 kgf m (413 lbf ft) and develops 180 bhp at 2 600 rev/min. The Henschel tradition of having direct-injection is followed and it has a minimum specific fuel consumption of 0·35 lb/bhp h (157 g/bhp h).

Whereas the 7·8 l. engine has two valves per cylinder, the 11·94 l. (729 in^3) engine has one inlet and two exhaust valves per cylinder. It develops its maximum power at 2 250 rev/min and has a maximum torque of 650 lbf ft (90 kgf m). The fuel consumption is good, the minimum specific figure being 0·342 lb/bhp h (154 g/bhp h). This engine has four-hole injectors whereas the smaller engine has three-hole injectors. Its weight is 815 kg (1 793 lb).

There are triple belts for driving the fan and air compressor and there are another two from the fan pulley for driving the alternator and water pump. A separate belt drive is provided off the crankshaft pulley for the power-steering pump when fitted. Both Henschel engines have wet cylinder liners and asbestos/steel sandwich gaskets. They have induction-hardened crank-shafts, the bearing material of the 7·8 l. engine being aluminium/tin whereas that of the 11·94 l. engine is copper/lead.

As the controlling interest in Henschel is now held by Daimler-Benz, some of the Henschel vehicles are now powered by the Daimler-Benz V10 and OM360 192 bhp six-cylinder diesels.

HINO (Hino Motors Ltd., 4, 2-Chome, Tori, Nihonbashi, Chuo-Ku, Tokyo, Japan)

Japan's motor industry is now very large, and one indication of the extent of its operations is the large range of diesel engines made by Hino. There are nine basic models, all except a new V8 being indirect-injection. They range from 4·3 l. to 15·96 l. (264 in^3 to 974 in^3) capacity.

All the in-line engines are six-cylinder designs except for an interesting horizontally opposed 12-cylinder engine, but even this is fundamentally two in-line sixes combined. Wet cylinder liners are used in all models. The general

The Hino V8 direct-injection diesel

specifications take a conventional form, with timing gears at the front, cast-iron block and head, two valves per cylinder and induction-hardened crankshaft.

When the 7·982 l. (486 in^3) model is 'doubled' in opposed 12-cylinder form its power is exactly doubled as well, becoming 320 bhp net at 2 400 rev/min, although the minimum specific fuel consumption creeps from 190 to 195 g/bhp h (0·42 to 0·43 lb/bhp h). The smallest Hino diesel is a 4·313 l. (264 in^3) model with the high governed speed of 3 200 rev/min. Its power is 90 bhp net.

The other six-cylinder models are a 7·014 l. (428 in^3) developing 140 bhp at 2 500 rev/min, a 9·036 l. (552 in^3) of 175 bhp at 2 350 rev/min, a 10·857 l. (663 in^3) of 175 bhp at 2 000 rev/min, and a horizontal of 10·178 l. (620 in^3) developing 195 bhp at 2 300 rev/min. This last engine is also available in turbocharged form, when it develops 230 bhp at 2 300 rev/min, maximum torque being boosted from 492 to 586 lbf ft (68 to 81 kgf m).

The direct-injection V8 is oversquare, the bore being 140 mm and the stroke 110 mm (5·51 in × 4·33 in) giving 13·55 l. (825 in^3). It develops 280 bhp at 2 600 rev/min and has a maximum torque of 608 lbf ft fairly well up the rev-range, at 1 800 rev/min. Its fuel consumption is fairly heavy for a direct-injection unit, being 0·41 lb/bhp h (187 g/bhp h). The V8's weight is 2 530 lb, which is almost the same as that of the 10·857 l. in-line six-cylinder model of 175 bhp.

Internally, the V8 breaks from previous Hino practice in having four valves per cylinder, steel gaskets, aluminium/tin (rather than copper/lead) bearings and eight-hole injectors.

INTERNATIONAL (International Harvester Company, 401 North Michigan Avenue, Chicago, Illinois, USA)

All three of a new series of diesel engines now made by International are V8s. The 7·55 l. (460 in^3) DV-462 and the 9 l. (550 in^3) DV-550 are allied designs —one being a larger bore version of the other. Both have a stroke of 4·312 in (109 mm). The DV-462 (186 bhp, SAE) has a bore of 4·3 in (105 mm) and the DV-550 (210 bhp, SAE) has a bore of 4·5 in (114 mm), making it oversquare.

Both these V8s are high-revving units, governed at 3 200 rev/min, and the maximum piston speed of 2 300 ft/min (11·6 m/sec) is high for a diesel engine. Also, maximum torque is developed quite high in the speed range—2 100 rev/min in the case of the DV-462. The engines were designed to be produced with much the same tooling as was already being employed to make V8 petrol engines.

Direct-injection of the German M.A.N. M-series type is used, in which a spherical open chamber in the piston crown has fuel sprayed directly on its wall by an inclined single-hole fuel injector. A high rate of air swirl is depended on for obtaining thorough mixing of the air, and is provided by the shape of the inlet port. For cold-starting, however, International discovered that a high rate of swirl was a bad thing. Much shorter cold-starting times are claimed to have resulted from arranging special flap valves in the air inlets so that air swirl is broken. For cold starting below 15° F (—9·4° C) ether injection into the induction manifold is advised, and International has devised a remote control system for ether injection.

Through having to use existing tooling, which happened to control the top-deck height of the cylinder blocks, International had to adopt pistons with shallow hemispherical-dome crowns. This involved inclined valves with 15 degrees between them, but has been useful in leaving more space for freedom in port and water-jacket design. A complication resulting from the tilted valves has been that the valve-rocker lips need a flat sliding face to contact the valve cap as well as a ball-ended valve stem. Positive rotators are fitted to the valves and the rockers are of diecast aluminium.

An in-line fuel injection pump made by the Holley Carburetter Company is used, this incorporating automatic advance/retard mechanism. A normal pattern of centrifugal governor is supplemented by an air-pressure diaphragm control for constant intermediate speed settings when using a power take-off.

Transverse tie bolts across the main bearing caps of the crankcase, load the cast iron in this area in compression, improving fatigue strength. The lubrication system of the larger V8 includes a jet at the bottom of each cylinder (which is not linered) to squirt cooling oil on the underside of the pistons.

The other V8 International weighs just short of a ton (2 220 lb or 1 010 kg) and is different in many respects from the two just described. For one thing it is

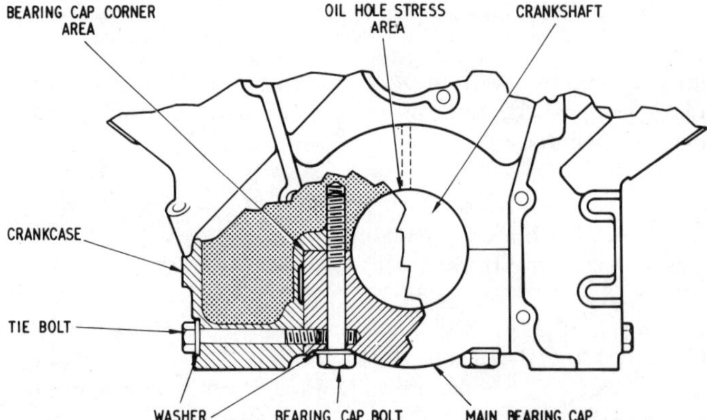

Main bearing caps of the International V8 are integrated into the crankcase structure by transverse tie bolts

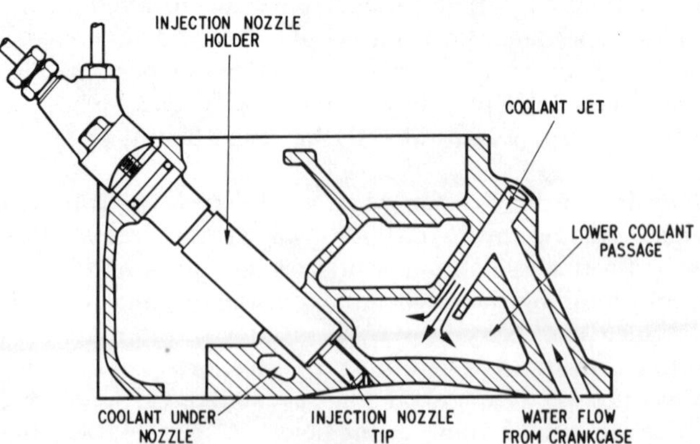

In the cooling system of the International DVT-573, extra water velocity is induced above the combustion chamber by top jets from a water gallery

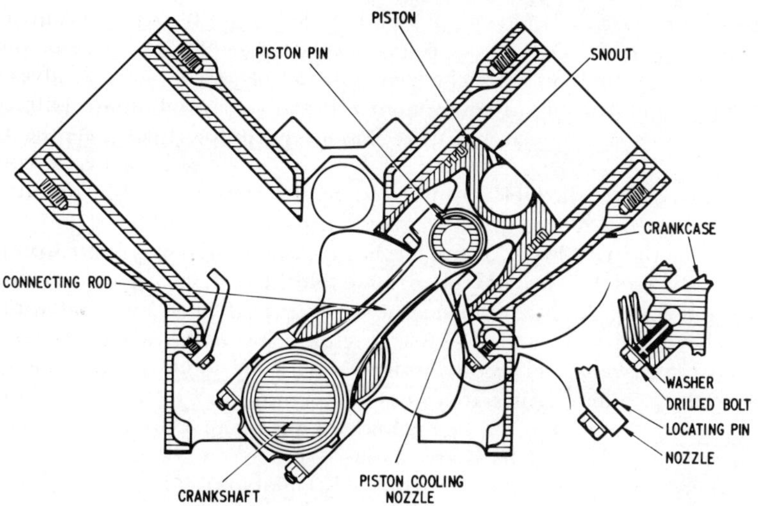

Section through the block of the International V8 showing the piston-cooling U *oil-jet arrangement*

turbocharged; for another it is bigger, having a capacity of 9·4 l. (573 in^3). Maximum power on the SAE basis is 260 bhp at 2 600 rev/min. Peak torque is said to be 578 lbf ft (80 kgf m) at 1 800 rev/min. Square bore/stroke proportions have been adopted, both being 4·5 in (114 mm).

Unlike the other International V8s, this turbocharged design, called the DVT-573, has wet cylinder liners and an open toroidal piston cavity. It uses a distributor fuel injection pump instead of the in-line pattern, and this incorporates an electric shut-off. Although the block and head castings are of cast iron, the main bearing caps are of ductile nodular iron, clamped by four ¾ in (19 mm) diameter set-bolts.

Much attention has been paid to the layout of the cooling system. Surrounding each wet cylinder liner is a horizontal baffle, having a small space between itself and the liner so that the water flow at the mid-section is accelerated and pressurised. This is designed to overcome the common problem of liner erosion. Water passing from the block to the head is channelled through jets which direct the flow first on to the bottom deck of the head.

ISUZU (Isuzu Motors Ltd., 22-10 Minami-oi, 6-chome, Shinagawa-ku, Tokyo, Japan)

Air cell pre-combustion chambers are used on most Isuzu engines. Only one model, the D920, is direct-injection. This engine also breaks with Isuzu tradition in another fundamental fashion—it has the same bore and stroke of 125 mm (4·93 in), which has allowed a higher governed speed for this 9·2 l. (561 in^3) six-cylinder; the speeds of the other large engines are limited by their stroke of nearly 6 in actually 5·91 in (150 mm).

Of the more popular Isuzu indirect-injection diesels, one range is based on a 130 mm (5·12 in) stroke; the other on a 150 mm (5·91 in) stroke. The one with the smaller stroke is available with either a 100 mm (3·94 in) bore, as the four-cylinder DA220 and six-cylinder DA120, or a 102 mm (4·02 in) bore, known as the model DA640—a six-cylinder of 6·4 l. (389 in^3) available either naturally aspirated or turbocharged with gross power outputs of 110 or 130 bhp. Higher ratings are often quoted for vehicle installations. High compression ratios—22 : 1—are used. A Bosch-type in-line fuel injection pump is used.

The big engines, the DH100 and E110, have a 150 mm (5·91 in) stroke. The DH100's bore of 120 mm (4·72 in) gives a capacity of 10·18 l. (621 in^3), while the bore of the E110 is 125 mm (4·93 in), which increases the capacity to 11·05 l. (674 in^3). The DH100 is available in turbocharged form and in horizontal form. A 'premium engine' concept has been followed with the bigger bore E110; as a high-output engine it has features such as sodium-cooled exhaust valves, oil-cooled pistons and a water-oil heat exchanger.

With the smaller engines, it is a bore of 83 mm (3·72 in) which tends to be the common dimension. A two-cylinder and a four-cylinder version of a 102 mm (4·02 in) stroke basis are made—1·1 l. (67 in^3) and 2·2 l. (134 in^3). Then there is a 92 mm (3·62 in) stroke edition of four-cylinders and a 1·99 l. (121·5 in^3) capacity. All these engines are indirect-injection.

LANCIA (Lancia & C. Fabbrica Automobili, Turin, Italy)

Aluminium-alloy for cylinder block and cylinder heads is one of the unusual features of the related 9·47 l. (578 in^3) and 10·52 l. (643 in^3) in-line six-cylinder four-stroke diesels made by Lancia. Heavy sections are employed in the light castings and the result is increased robustness rather than great weight saving, the 10·52 l. engine weighing 1 984 lb (900 kg).

Four valves to each cylinder give good breathing so that the power outputs are quite high. The 9·47 l. engine develops 176 bhp at 2 300 rev/min and the

The Lancia in-line six-cylinder engine, which has aluminium alloy block and heads

10·52 l. model, 209 bhp at 2 200 rev/min. Both engines have a 122 mm bore (4·80 in), but whereas the smaller engine has a 135 mm (5·31 in) stroke, the 10·52 l. unit has a stroke of 150 mm (5·91 in). Four-hole injectors spray fuel into toroidal piston cavities. Wet cylinder liners are used for both engines.

The vee-twin air compressor is gear driven and is mounted on the front of the timing case. Otherwise, the auxiliaries are belt-driven. There are three sets of twin vee-belts for the fan/water-pump, the generator, and the power-steering oil pump. The plate for the front mounting of the engine is sandwiched between the cylinder block and the timing case. The 5 in (127 mm) starter motor is flange-mounted in a high position on the flywheel housing.

Careful attention has been paid to oil filtration, there being a Glacier centrifugal bypass filter in addition to a full-flow filter connected to an oil-water heat exchanger with a thermostatic valve.

The performance characteristics of the Lancia engines are such that high torque is developed at low speeds—the maxima, of 458 lbf ft (63·3 kgf m) for the 9·47 l. and 555 lbf ft (76·7 kgf m) in the case of the 10·52 l., are developed at 900 rev/min. But the best fuel economy is given in the upper half of the rev-range, the minimum being at 1 600 rev/min. Lancia is now controlled by Fiat.

LEYLAND [British Leyland Motor Corporation, Leyland, Lancs, England: Leyland Motors (Scotland) Ltd., Bathgate, Scotland]

Diesel engine production at Leyland began in 1931 with a 5·7 l. (347 in^3) four-cylinder and an 8·6 l. (524 in^3) six-cylinder overhead-camshaft design. Direct-injection was used but was of an exclusive pattern never copied. The piston cavity was of deep bucket shape and a single-hole injector sprayed fuel down one side of it. The result was remarkably smooth combustion—smoother

Top end of the original Leyland direct-injected engine showing the overhead camshaft and pot-cavity piston

even than most indirect-injection engines. There was some sacrifice in fuel economy; during the Second World War a push-rod 7·4 l. (451 in^3) six-cylinder engine was developed with large diameter toroidal piston cavities fed with atomised fuel by four-hole injectors. Though a smaller engine than the previous Leyland design (and much noisier) it developed 100 as against 92 bhp and set a high standard in fuel economy.

Many internal features of this 7·4 l. engine were incorporated in a new 9·8 l. (600 in^3) six-cylinder diesel introduced in 1946 and called the 600. This engine has formed the basis for Leyland engines for over 22 years. Originally it developed 125 bhp at 1 800 rev/min but its power output has been increased to 166 bhp at 2 200 rev/min over the years and an 'economy' version developing 140 bhp at 1 700 rev/min has been one of the most economical diesel engines in the world with a minimum specific fuel consumption bettering 0·34 lb/bhp h (154 g/bhp h).

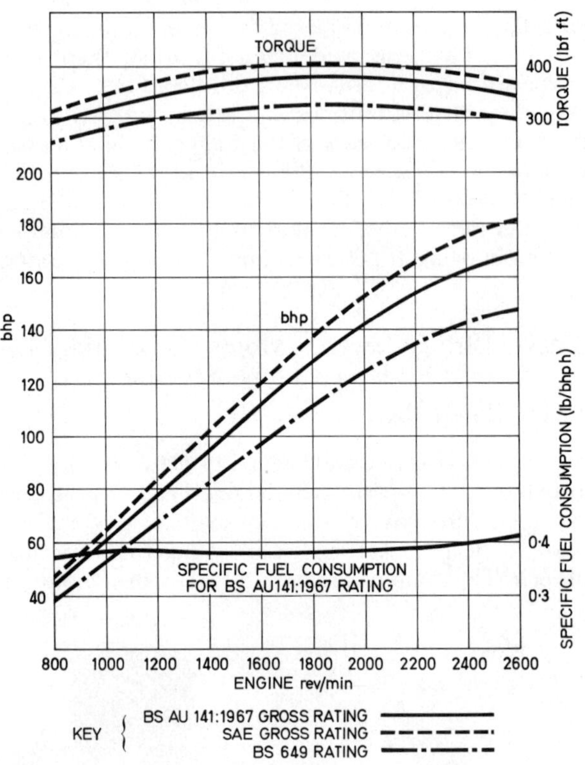

Performance curves of the normally aspirated Leyland 500 model

An 11·1 l. (677 in^3) version of this engine, with a larger bore and stroke, swelled the Leyland range a few years later and this 680, became more and more popular until it eclipsed the 600. Originally it developed 150 bhp at 2 000 rev/min but now is a 200 bhp engine when normally aspirated and generates 240 bhp when turbocharged.

The honour of having the highest quantity production of any one Leyland type went to a much smaller design, however. From a 5 l. (311 in^3) 75 bhp six-cylinder engine, called the 300, introduced in 1948 as a diminutive high-speed version of the successful 600, was developed first the 105 bhp 350, later the 110 bhp 375 and then revised versions in 6·07 l. and 6·54 l. form, called the 370, 400 and 401, these figures indicating in^3 capacity.

The 400 engine develops 126 bhp—as much as the original 600— and is still almost the same overall size as the 300 from which it is originally derived.

It has been, and still is, a popular engine in several other makes of vehicle apart from Leylands, as well as finding many industrial applications (as have indeed the other Leyland designs). The 401 is a higher-revving (2 600 against 2 400 rev/min) version of the 400 with rotary-swirl inlet ports. It develops 138 bhp.

All the Leyland engines are built in horizontal as well as vertical form and many permutations of specification are now available to suit all sorts of installations.

When Leyland went into mass-produced car manufacture by taking over Standard-Triumph it inherited the Standard four-cylinder 2·26 l. (137 in^3) indirect-injection engine and brought it into the Leyland range, changing the construction to dry liners and arranging oil-spray cooling of the pistons. It was not long before a larger capacity version, termed the OE160, was developed. Both bore and stroke were increased, making the capacity 2·61 l. (159 in^3). This engine developed 60 bhp against 52 bhp of the OE138, as the Standard diesel became known. The governed speed of these engines was 3 000 rev/min. Their timing drive was by chain, and they had distributor fuel injection pumps. Except for a few horizontal 600 engines the Leyland company does not otherwise use distributor pumps. Leyland was a pioneer of nitride-hardening of crankshafts in which the forged-steel is put in a state of compressive stress,

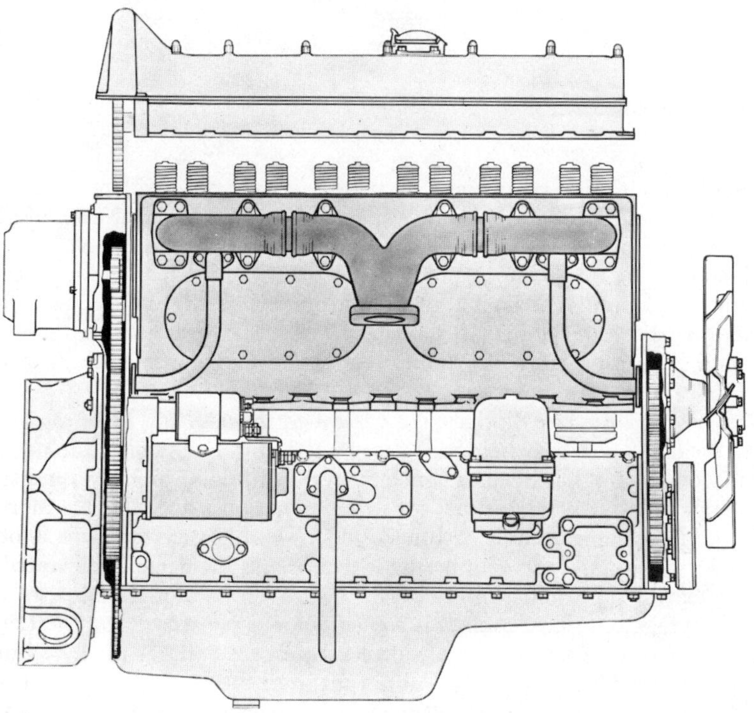

The Leyland 500 design has the cylinder head and block cast in one piece, the crankcase being separate. An overhead camshaft is used, driven by a train of spur gears from the flywheel end of the crankshaft

greatly enhancing fatigue strength. Aluminium/tin bearings have now replaced the copper/lead ones which were at one time almost universal. All Leyland engines have dry cylinder liners.

The Leyland company joined the high-power stakes a few years ago as a result of taking part in a multi-fuel diesel engine development programme sponsored by the British Government for military purposes. The 700 bhp

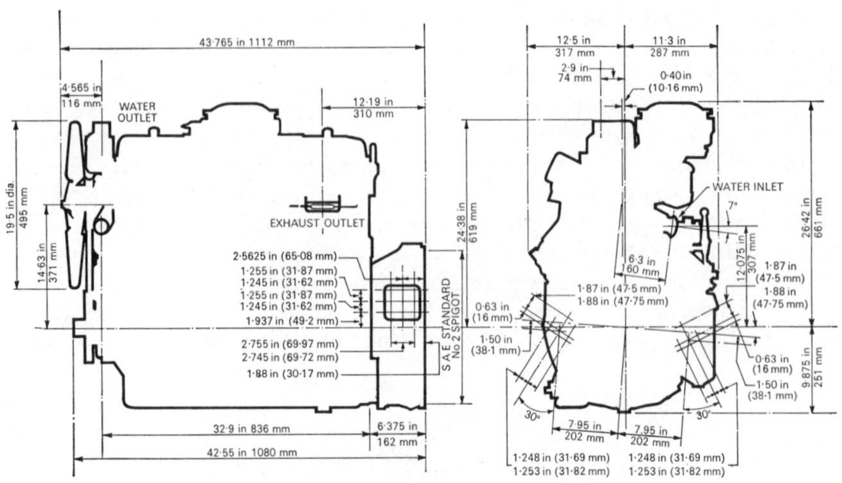

Compact installation dimensions of the 6·54 l. (399 in³) 138 bhp Leyland 401

twin-crankshaft opposed-piston 19 l. two-stroke diesel which emerged is now on general sale. It is an inherently expensive engine but packs a lot of power into compact dimensions. Its weight is less than 2 tons.

Now there is a drive among the Leyland engine design team for high power/weight ratios and a turbocharged replacement for the 600-based series of engines has been developed. It has several novel features and promises to be the start of another era of Leyland diesel engines with a long and successful production run.

The new in-line six-cylinder engine is called the 500, and has a capacity of 8·2 l., the bore and stroke being 118 mm by 125 mm. It is available in normally aspirated form but has been designed from the outset for high-pressure turbocharging. This is why an unusual construction has been adopted: the head and cylinder block being a single casting, while the crankcase is separate. This has, of course, eliminated the cylinder head gasket and the decking between head and block which normally interferes with water circulation and heat transfer. The toroidal-cavity pistons are oil-cooled.

By using an overhead camshaft, driven by a gear-train at the flywheel end of the engine, valve-bounce does not occur until the speed has been raised to 3 750 rev/min, allowing plenty of margin for overspeeding on overrun above the governed 2 600 rev/min. Moreover, the absence of push-rods, as well as cylinder head studs, has given much more freedom in port layout. The result is that very high power outputs are obtainable. In normally aspirated form the 500 develops 170 bhp (BS) but when turbocharged with intercooling there is potential for up to 400 bhp. The initial top rating of the turbocharged 500 is, however, 260 bhp.

The aim has been to design a minimum life of 300 000 miles (483 000 km) between major overhauls. High quality valve steels, chromium-plated dry cylinder liners and elimination of belt drives all help to this end.

The 500 engine has metric dimensions and threads and is designed to have a versatile layout to suit different vehicle, industrial and marine applications. It will run in either direction, the drive can be taken from either end of the crankshaft, the auxiliaries can be on either side and so can the manifolds. The whole engine can be supplied in either vertical or horizontal form and will work satisfactorily at an angle of 45 degrees in either plane when oil pumps are fitted at each end of the sump. Its weight is 753 kg (1 659 lb).

Being now part of the British Leyland Motor Corporation, the diesel engines which used to be marketed by Austin, Morris or BMC are now known

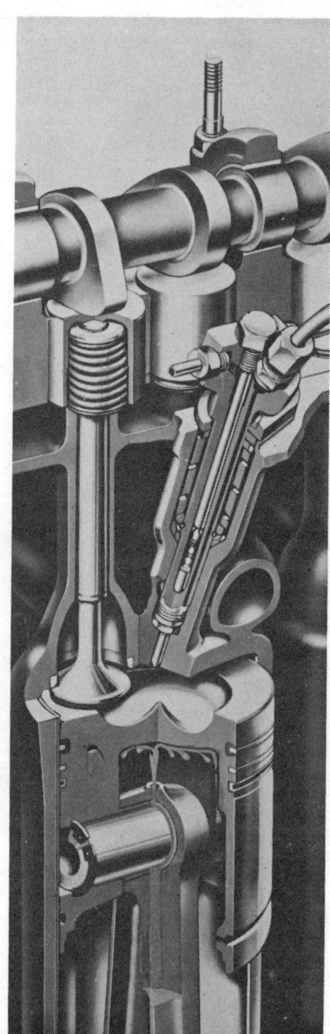

Section through the mono-head and overhead camshaft of the Leyland 500 engine

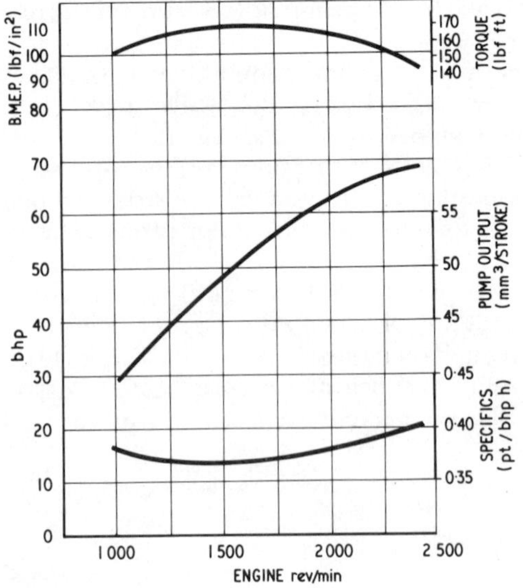

Performance curves for the 3·8 l. (231·8 in^3) Scottish Leyland engine

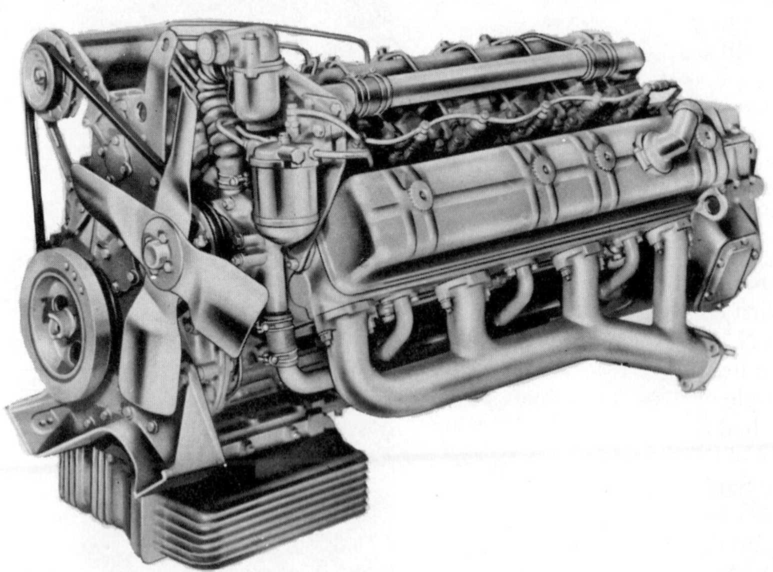

The 5·7 l. (345·3 in^3) six-cylinder Scottish Leyland engine is canted so that it fits under the floor of a truck cab

as Leylands. In the late 1940s, Morris had already seen the need for a small diesel and introduced a 4·5 l. (274 in^3) six-cylinder built under Saurer licence. Further demand to extend the scope of application of diesels to goods vehicles coincided with the formation of the British Motor Corporation (the Austin-Morris merger) and the company designed its own range of four- and six-cylinder engines in 1954. The first was a direct-injection 3·4 l. (207 in^3) four-cylinder. This is basically still in production, although with an increased bore of 100 mm (3·94 in) it gives a capacity of 3·8 l. (231·8 in^3), power increasing from 58 to 68 bhp at 2 400 rev/min.

The 5·1 l. 311 in^3) six-cylinder engine which followed is still in highly successful production. A 100 mm (3·94 in) bore version of this engine also has been introduced, making the capacity in this case 5·7 l. (345·3 in^3), and increasing the power from 90 to 105 bhp at 2 400 rev/min. All these engines have simple bath-tub piston cavities, wet liners and, now, timing drive by gears. The four-cylinder engine has an in-line fuel pump but the six-cylinder engines now have distributor pumps.

Since Leyland took control of the diesel side, important improvements have been made to the 5·7 l. engine—mainly to its cooling system. The longitudinal water gallery, cored along one side of the upper end of the cylinder block, now feeds the cylinder jackets instead of the cylinder head, as formerly. This is claimed to have resulted in such uniform temperature distribution that there is only 1° F (0·55° C) difference between the hottest and coolest regions.

BLMC was quick to recognise the potential of diesels in light vans and taxis. Its first venture in this field was the indirect-injection 2·2 l. (134 in^3) four-cylinder engine; a great success it is still strongly in production. It develops 55 bhp at 3 500 rev/min and is now fitted with a distributor pump. The timing drive is by chain in this case.

Few people expected a diesel engine any smaller than this to be an economic proposition. Still less did they expect a diesel sharing components with a petrol engine to be reliable. But BLMC proved the sceptics wrong. By producing a 1·5 l. (91·5 in^3) diesel, largely with existing tooling for a petrol engine of the same size, they were able to offer a diesel alternative for a price untouched up till that time. It is offered as an option for light vehicles.

M.A.N. (Maschinenfabrik Augsburg-Nürnberg A.G., Abholfach, 8500 Nürnberg 2, W. Germany)

Before automotive engines were hardly heard of the M.A.N. company built the first engine to Dr. Rudolf Diesel's design in 1897. It is not surprising that M.A.N. is proud of being in on the ground-floor of diesel evolution. It has continued to play an important part in diesel development ever since. For example, it has been a champion of direct-injection from the earliest days. It even had a primitive form of toroidal combustion chamber before the significance of this shape was generally appreciated and properly developed by the Swiss Saurer concern.

After trying out a two-stage combustion system compromising between air cell and direct-injection layouts a radical type of direct-injection was devised by Dr. Meurer. Termed the M-type combustion system this employed a pure spherical chamber in the piston crown, with a constricted opening. This combustion system is a prime feature of present-day M.A.N. diesel engines and has been developed to a high state of efficiency while retaining a pressure-rise characteristic smoother than is normal with direct-injection.

With the M-type spherical-chamber system, fuel is sprayed deliberately on the wall of the combustion chambers. Vaporisation, a very high rate of rotary swirl induced by special inlet-port design and the squish into the chamber are relied on to pull the fuel steadily into the air stream. Minimum specific fuel consumptions in the 0·348 to 0·352 lb/bhp h (153 to 159 g/bhp h) range are obtained from the modern M.A.N. engines.

Four engines form the nucleus of the M.A.N. range of six-cylinder automotive engines. The smallest is a 5·275 l. (322 in^3) six-cylinder with a bore of 100 mm (3·94 in) and a stroke of 112 mm (4·41 in). It develops 126 bhp at 2 900 rev/min. Then comes a 7·25 l. (442 in^3), with a bore of 108 mm (4·26 in) and a stroke of 132 mm (5·2 in), developing 156 bhp at 2 500 rev/min; a 9·66 l. (589 in^3), with a bore of 121 mm (4·77 in) and a stroke of 140 mm (5·51 in), developing 186 bhp at 2 200 rev/min; and a 150 mm (5·91 in) stroke engine, normally aspirated with a 123 mm (4·84 in) bore to give 230 bhp from 10·689 l. (653 in^3), or with a 121 mm (4·77 in) bore to give 256 bhp from 10·344 l. (635 in^3) when turbocharged or 192 bhp when normally aspirated. A variation of the bigger six-cylinder engine is a bigger bore version with the power raised to 235 bhp.

For some time the M.A.N. range has consisted mainly of in-line six-cylinder

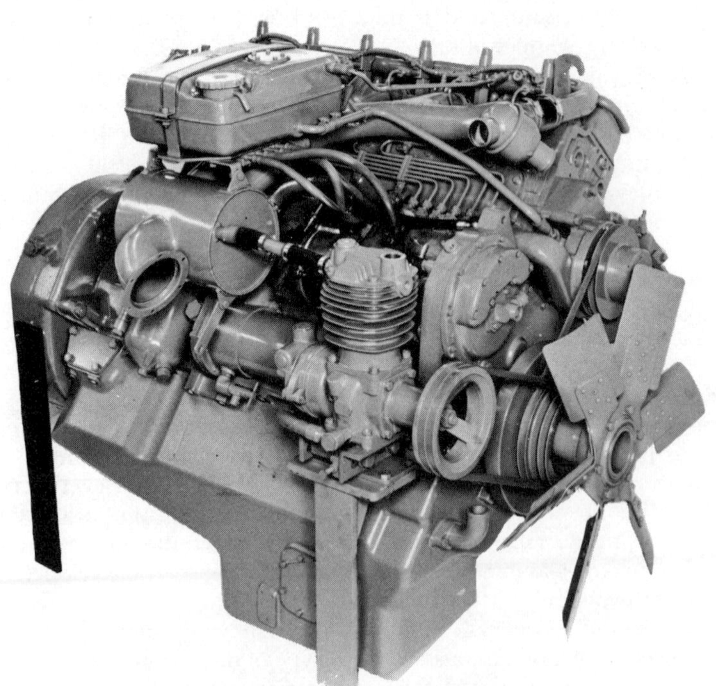

Canted version of the 10·35 l. (653 in^3) M.A.N. six-cylinder engine

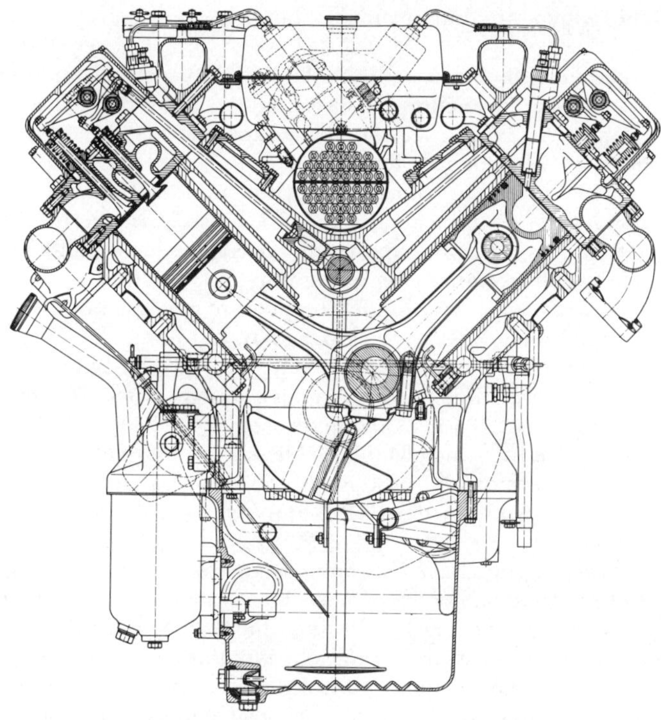

Cross-section through the M.A.N. 90-degree V8 diesel

units, but a V8 has now been introduced to meet the demand for a higher-powered engine compact enough not to take up much room in the forward-control cab of a truck.

The V8 had a capacity of 14·96 l. (912 in^3) when it was introduced in 1967, but it was enlarged to 15·45 l. (973 in^3) in 1969. Its bore is 128 mm (5·04 in) and its stroke, 150 mm (5·91 in). Maximum power is now 304 bhp at 2 200 rev/min and the peak torque 105 kgf m (761 lbf ft). Its weight is 2 340 lb (1 062 kg). A novel feature is the V8 fuel injection pump developed by Bosch for this engine. This pump's shortness (it is driven from the rear of the crankshaft) leaves room for a tubular oil/water heat exchanger in front of it between the banks of the vee. Inlet and exhaust are on the opposite sides of each head and there are four valves per cylinder. A turbocharged version of the V8 is under development, and is designed to deliver 350 bhp.

NISSAN (Nissan Diesel Motor Co. Ltd., Kawaguchi, Saitama, Japan)

Out of the 3 000 engines a month made by Nissan, half are automotive diesel engines. There are three-, four-, five and six-cylinder in-line models as well as a V8. These are mainly intended for the trucks, buses, concrete mixers and mobile cranes which the company makes, but many go into industrial applications.

The company is rather unusual in making both two-stroke and four-stroke engines. The two-stroke engines are based on a cylinder of 110 mm (4·34 in) bore and 130 mm (5·12 in) stroke. This gives capacities of 3·7 l. (226 in^3) for the three-cylinder, 4·94 l. (301 in^3) for the four-cylinder, 6·18 l. (377 in^3) for the five-cylinder, 7·41 l. (452 in^3) for the six-cylinder and 9·88 l. (601 in^3) for the V8. All these have a compression ratio of 16:1 and have overhead valves of 21–4NS material. As might be expected, these engines have a high power output, a typical specific weight being 3·6 kg/bhp (7·93 lb/bhp).

The three-, four- and five-cylinder engines are governed at 2 400 rev/min at which they produce respectively 130, 175 and 215 bhp (DIN), with maximum torques of 45·5, 63 and 75 kgf m (328, 456 and 542 lbf ft) at 1 400 rev/min. The in-line six-cylinder and V8 models are governed at 2 200 rev/min at which they produce 240 and 330 bhp respectively. Their maximum torques are 92 and 122 kgf m (665 and 884 lbf ft). These latter two engines have slightly heavier fuel consumption than those with fewer cylinders; i.e. 190 g/bhp h compared with 185 g/bhp h (0·418 lb/bhp h against 0·407 lb/bhp h) for the smaller engines.

The two-stroke engines have chromium-plated wet cylinder liners, whereas the four-stroke engines do not have liners. The two-stroke engines have pistons of malleable cast iron in order to withstand the high mechanical and thermal stresses imposed by two-stroke operation, but the four-stroke engines have normal aluminium-alloy pistons. In both cases there is an unusually large number of piston rings, the two-stroke pistons having six and the four-stroke pistons, five rings. The oil-control ring of the two-stroke engines has an expander behind it.

Direct-injection is used for the two-stroke engines, but the four-stroke engines are indirect-injection with swirl chambers in the heads. There are three models of four-stroke engine. Two of them are based on the same cylinder of 83 mm (3·27 in) bore and 100 mm (3.94 in) stroke but having four and six cylinders in line. These engines are at the smaller end of the Nissan scale. The four-cylinder engine has a capacity of 2·16 l. (132 in^3) and the six-cylinder engine a capacity of 3·25 l. (198 in^3) which is unusually small for a six-cylinder diesel engine. The third four-stroke Nissan model is a four-cylinder engine with the same bore as the others but a shorter stroke—92 mm (3·62 in). This gives it a capacity of 1·99 l. (121 in^3).

All three engines have a compression ratio of 22:1 and a minimum specific fuel consumption of 195 g/bhp h (0·429 lb/bhp h). The smallest engine develops 60 bhp at 4 000 rev/min with a torque of 13 kgf m (94 lbf ft). The bigger four-cylinder, long-stroke engine develops 65 bhp and the six-cylinder version 98 bhp, both at 4 000 rev/min. The corresponding maximum torques of the long-stroke engines are 14·5 and 22 kgf m (105 and 159 lbf ft). In all cases the maximum torque is developed at 1 800 rev/min. These engines weigh 187, 190 and 268 kg (411, 417 and 589 lb), in ascending order.

The four-stroke engines have 21–4NS overhead valves operated through flat tappets, whereas the two-stroke engines have roller tappets. Like all the Nissan engines the timing drive is by gears. The fuel injection equipment is of the normal jerk pump layout and incorporates a centrifugal automatic advance/retard mechanism in the drive.

ÖAF (Österreichische Automobil-Fabriks A.G., Brunnerstrasse 72, Vienna 21, Austria)

Two automotive engines are produced by Ö.A.F. They are in-line six-cylinder direct-injection designs. The smaller, 9·5 l. (580 in^3) engine has a bore of 120 mm (4·73 in) and a stroke of 140 mm (5·51 in). It develops 180 bhp (DIN) at 2 100 rev/min and has a maximum torque of 478 lbf ft (66 kgf m) developed at 1 400 rev/min. The other engine has a 121 mm (4·76 in) bore and a stroke of 150 mm (5·9 in). These give a capacity of 10·35 l. (631 in^3), from which 215 bhp is obtained at 2 200 rev/min. Its maximum torque is 76 kgf m (550 lbf ft).

A composition gasket is used between the cast-iron heads and block, and dry cylinder liners are fitted. The timing drive is by gears off the front end of the crankshaft, which is induction hardened. For several goods vehicles made by Ö.A.F., Leyland diesel engines are installed.

PEGASO (Empresa Nacional de Autocamiones SA, Lagasca 88, Madrid 6, Spain)

There is useful cross-feeding of technical data between Pegaso in Spain and the Leyland company in Britain, and one result of this is a 6·54 l. (399 in^3) Pegaso six-cylinder in-line engine which closely follows the design of the Leyland 400 engine and has the same power output (136 bhp gross, on the SAE basis). Detail differences include special attention to the cylinder-liner treatment and the use of aluminium/tin rather than copper/lead bearings. A four-cylinder version of this engine is also made, having the same bore and stroke. The capacity of this engine is 4·3 l. (267 in^3) and it develops 95 bhp (SAE).

The bigger Pegaso in-line six-cylinder engines bear no direct relationship with Leyland designs, however. Mainly as a result of having aluminium-alloy cylinder blocks (though cast-iron heads) they are comparatively light. The latest 10·52 l. (646 in^3) Pegaso engine weighs 1 440 lb (653 kg), it has wet cylinder liners and direct-injection with toroidal combustion chambers is used. The bore is 120 mm and the stroke 155 mm (4·72 × 6·1 in), giving an SAE power output of 205 bhp. This engine is a slightly larger bore version of the 10·17 l. (621 in^3) six-cylinder engine which develops 185 bhp. Now here is a 260 bhp turbocharged edition of the 10·52 l. engine and there is a new 11·94 l. (729 in^3) turbocharged in-line 'six' for which the high power of 352 bhp (SAE) is claimed.

PERKINS (Perkins Engines Ltd., Peterborough, England)

Perkins makes over 350 000 engines a year, nearly 250 000 of which are made in the English factory. Other factories in Argentina, Bulgaria, France, Germany, India, Japan, Mexico, Spain and Turkey are expanding. High demand for the smaller diesel engines in which Perkins specialises stimulates such quantity production. Four-, six- and V8 cylinder engines from 20 to 205 bhp are in the Perkins automotive range.

For many years a traditional feature of Perkins designs was the Aeroflow combustion chamber which took the form of a large-throated air cell, giving a useful compromise between indirect- and direct-injection. This combustion system has now been almost completely superseded by more orthodox combustion chambers. Pure indirect-injection has been introduced for the small four-cylinder diesels, while for the larger engines, direct-injection with toroidal piston cavities is used.

The smallest four-cylinder Perkins automotive engine was at first the four-cylinder 1·62 l. (98 in^3) 4.99. This had a governed speed of 3 600 rev/min, but for car and taxi installations was governed at 4 000 rev/min. It weighed

The Perkins four-cylinder 4.236

only 380 lb (172 kg). A later development of this has been the 4.108, which has the same 3·5 in (89 mm) stroke as the 4.99 but a bore of 3·125 in (79 mm) against 3 in (76 mm). This increases the capacity to 1·76 l. (108 in^3) and its power to 52 bhp. Attention has been paid to gasket and cylinder head design and it differs from the 4.99 by having dry instead of wet cylinder liners and a tufftrided crankshaft. The pistons have their top rings seated in steel inserts.

Now there is another small four-cylinder Perkins—the 2·52 l. (154 in^3) 4.154, which has a bore of 3·5 in (89 mm) and a stroke of 4·0 in (102 mm). It delivers 68 bhp (DIN) at 3 600 rev/min. The 3·33 l. (203 in^3) engine—the 4.203—is now one of the oldest Perkins designs with the almost traditional Perkins 5 in (127 mm) stroke. This alone has the original Aeroflow combustion chamber.

The first direct-injection Perkins engine was the 5·8 l. (354 in^3) 6.354, which was a highly successful replacement for the previous similarly sized indirect-injection engine called the R6. Specific fuel consumption improved from about 0·42 to 0·385 lb/bhp h (190 to 174 g/bhp h) with no sacrifice in power. In its latest form the fuel consumption has been improved to 0·37 lb/bhp h (167 g/bhp h). The 6.354 led the way for higher-revving diesel engines and showed that a 2 800 rev/min governed speed was a practical proposition for direct-injection engines exceeding 5 l. (305 in^3), capacity, without reliability being eroded. Its net maximum power is 118 bhp. A vertical distributor fuel injection pump is used in conjunction with four-hole injectors, and dry cylinder liners are fitted. The 6.354 is available in either vertical or near-horizontal forms and the water pump can be positioned on the front of either the block or the head to suit different installation requirements. A heavy-duty version is offered with such features as a nitrided crankshaft, armoured top ring-grooves, more oil capacity and a large overspeed reserve.

For industrial and marine applications there is a turbocharged version of the 6.354 developing a maximum gross power of 160 bhp at 2 500 rev/min. For automotive duty the peak temperatures and pressures in the turbocharged 6.354 have been reduced by employing an air-to-air intercooler after the turbocharger. This has also produced better utilisation of air and has improved fuel consumption. The DIN net power is 143 bhp at 2 600 rev/min, and the maximum torque, 358 lbf ft (49·5 kgf m).

It was the 6.354, in a mechanically supercharged form which was used by Perkins for the development of its differentially-supercharged diesel engine (termed D.D.E. for short) power-transmission pack. In this exclusive system

The 9·9 l. (605 in^3) Perkins V8.605

the engine drove a torque converter via an epicyclic differential gear in which the reaction was taken by the shaft driving the supercharger. The greater the load imposed by the torque converter the faster the supercharger was driven. The result was that a very high degree of supercharge was given at low crankshaft speeds and the engine developed almost constant horsepower over its speed range. Combined with the torque multiplication of the special torque converter the equivalent of a 7:1 ratio-spread transmission was provided, yet without any gear changing except when wanting to travel in reverse. Development of the D.D.E. system has been shelved indefinitely.

A four-cylinder version of the direct-injection 6.354 is now produced. This is called the 4.236 and has the same mechanical features except that the distributor fuel pump is driven horizontally. Its capacity is 3·86 l. and its maximum power 80 bhp at 2 800 rev/min. It is the most economical Perkins engine, having a specific fuel consumption of 0·36 lb/bhp h (163 g/bhp h).

The latest Perkins engines are V8s which extend the range of engine sizes farther than before. The V8.510 has a capacity of 8·36 l. (510 in^3) and has bore-stroke proportions more square than any Perkins made hitherto. Its bore is 4·25 in (108 mm) and its stroke 4·5 in (114 mm). Its DIN net power is 167 bhp and its maximum torque is 380 lbf ft (52·5 kgf m). The larger 9·9 l. V8.605 is the same size outwardly but has a longer stroke. This V8 develops 205 bhp. It has a French Sigma rotary fuel pump and Italian Omap injectors.

Like the 6.354, two V8s have direct-injection toroidal combustion chambers and machined inlet porting. An in-line fuel injection pump on the V8.510 has an automatic timing mechanism, while the injectors are on the outside of the vee for easy servicing. The overall dimensions of the Perkins V8, only 36 in long, 33 in wide and 32·5 in high (914 × 838 × 826 mm), make it very compact. This helps keep the weight down to 1 440 lb (653 kg).

Most of the Perkins automotive engines are available in marine and industrial form, but there are a few engines designed expressly for industrial or agricultural use. These are the three-cylinder P3.144 (2·36 l.) and 3.152 (2·5 l.), and the four-cylinder 4.270 (4·42 l.) and 4.300 (4·95 l.); respective capacity equivalents are 144, 154, 265 and 302 in^3.

ROLLS-ROYCE (Rolls-Royce Ltd., Oil Engine Division, Shrewsbury, England)

After several years of offering a wide range of diesel engines embracing four-, six- and eight-cylinder units, Rolls-Royce rationalised its automotive designs in 1966 by developing a single basic six-cylinder engine with a 5·125 in (130 mm) bore and 6 in (152 mm) stroke. From this one size of engine four models have been derived, all being marketed under the name Eagle. The capacity is 12·17 l. (743 in^3).

There are two naturally aspirated Rolls-Royce Eagles and two turbocharged. One of the naturally aspirated models is rated at 205 bhp (BS) and has a maximum torque of 559 lbf ft (77 kgf m) at 1 400 rev/min—the Eagle 205. The partner to this is the Eagle 220, which has a higher rating, developing 220 bhp (BS) at the same governed speed of 2 100 rev/min and with a maximum torque of 605 lbf ft (83 kgf m) at 1 400 rev/min. Both engines weigh 2 410 lb (1 092 kg). A minimum specific fuel consumption of 0·369 lb/bhp h (167 g/bhp h) is achieved by the 205, and 0·376 lb/bhp h (170 g/bhp h) by the 220.

The turbocharged versions are known as the Eagle 265 and 305, which have a compression ratio of 14 : 1 instead of 16 : 1. The 265 develops a maximum power of 265 bhp (BS) at 2 100 rev/min and a torque of 755 lbf ft (104 kgf m) at 1 400 rev/min. The higher rated Eagle develops 305 bhp and has a torque of 855 lbf ft (118 kgf m). The specific fuel consumptions are considerably better than for the naturally aspirated models, being 0·355 lb/bhp h (163 g/bhp h) for the 265 and 0·348 lb/bhp h (158 g/bhp h) for the 305. The turbocharged engines weigh nearly 100 lb (45 kg) more than the supercharged versions.

Apart from the various power settings, the Eagle engines are very similar to one another. They are all direct-injection designs with toroidal piston cavities. They have wet cylinder liners, two valves per cylinder (exhaust valves are sodium cooled), timing gears at the front end, zinc-plated steel gaskets, cast-iron block and heads, lead/bronze bearings and nitrided crankshaft.

In July 1970 Rolls-Royce introduced its Eagle Mark II range of engines. These are basically the same as the earlier Eagle units but have detail modifications so that they can develop maximum power in accordance with previously published figures yet satisfy the test conditions laid down in British Standard AU141:1967.

The modifications were principally to gain more efficient combustion and reduce internal friction. Valve ports were reshaped, a swirl-inducing crescent machined beside each inlet-valve seat, valve heads recessed in the cylinder

The Rolls-Royce Eagle 12 l. (743 in³) engine, developing, on British Standard rating, either 205 or 220 bhp

head to eliminate cut-outs in the pistons, and the exhaust manifold redesigned for coupling to a 4 in (101·6 mm) bore pipe. Pistons were reprofiled, with relieved skirts to minimise drag, and a new three-ring layout adopted in which the two lower rings are chromium plated. The top compression ring has a molybdenum inlay. The second compression ring is of L-section, pistons are carefully matched and great care is taken to equalise the power output of all six cylinders. The overspeed capability of the valve gear was raised by making it stiffer.

In the Mk II development some weight was saved by simplifying the air compressor and fuel injection pump mounting. The air compressor (now water cooled instead of air cooled) is now flange-mounted from the back of the timing case instead of on a separate saddle. The injection pump is driven through a large flexible coupling from the back of the compressor. Both units are pressure-fed with oil from the engine's lubrication system. Readier accessibility for servicing has been provided by repositioning the oil filters low down at the front end. This makes them accessible from beneath, without having to raise engine covers and without risk of contaminating the outside of the engine with spilt oil.

Fan-drive reliability has been ensured by using quadruple vee belts. They have more than adequate capability to drive a hydraulic pump for power-assisted steering systems, as well as the fan and water pump. Using a multi-bladed fan of large diameter and having minimum obstruction to air-flow behind it have all contributed to an unusually low noise level from this source. Fan noise is often a second-rate contributor to an engine's total noisiness, says Rolls-Royce.

ROVER (The Rover Co. Ltd., Solihull, Warwickshire, England) (part of British Leyland Motor Corporation)

When Rover decided to offer a diesel alternative for the Land Rover cross-country 1 ton four-wheel-drive vehicles, it took a robust petrol engine as the basis so that use could be made of much existing tooling. The result has been a four-cylinder engine which is oversquare, developing 67 bhp at 4 000 rev/min.

Its capacity is 2·28 l. (140 in^3), bore 3·562 in (90·5 mm), stroke 3·5 in (89 mm). A maximum torque of 105 lbf ft (14·5 kgf m) is generated at 1 800 rev/min. It weighs 533 lb (241 kg). The minimum specific fuel consumption is 0·51 lb/bhp h (231 g/bhp h). Chain drive is used for the timing, and the crankshaft, which is unhardened, has three copper/lead main bearings. Dry cylinder liners are fitted.

Driven from a skew gear on the camshaft is a vertical CAV distributor pump

The Rover 2·28 l. (140 in^3) four-cylinder diesel with Land Rover dual gearbox

serving the combined conical-spray and auxiliary-hole injectors which are clamped in the cylinder head. The injectors spray into Ricardo Mk V lantern-section combustion chambers in the head, communicating with the cylinders by narrow throats. This indirect-injection system gives thorough combustion even at high engine speeds. To cope with flood-water conditions when installed in Land Rovers, the crankcase breather is extended to above the rocker cover.

SAURER (Société Anonyme Adolphe Saurer, Arbon, Switzerland)

It may, perhaps, truly be said that the Saurer Company was largely responsible for the rapid growth of the compression-ignition engine in road transport, this company having put such engines on a production basis in 1928. For a number of years, Saurer diesel engines were manufactured under licence by Armstrong Saurer, at Newcastle-upon-Tyne (production of which ceased in 1937) and by Morris Commercial before it was merged with Austin to form the British Motor Corporation (now British Leyland).

A sectioned Saurer engine, as made under licence by O.M. of Italy, showing the drive to the screw-lobe supercharger and the oil–water heat exchanger below it

Originally designed around the Acro pre-combustion system, the Saurer engine was modified from time to time until, in November 1934, the dual-turbulence toroidal direct-injection design was introduced. Improved volumetric efficiency and central location of the injector were obtained by adopting four valves per cylinder. These were operated by two pairs of rockers for each pair of valves, one set being actuated by vertical pushrods from a camshaft on one side of the engine, hardened steel cup ends engaging with ball-ended adjusting screws on the outer ends of the rockers. The inner ends of the rockers were fitted with hardened steel roller bearings on the inner ends of the other set of rockers, which swung from a rocker shaft on the other side of the power unit. Apart from the ball-and-socket adjustment on each push-rod, only one screw was required to adjust both valves simultaneously. This arrangement is still used on some Saurer engines.

Following established Saurer practice, these improved engines were of monobloc construction and had wet cylinder liners. Three cylinder sizes and either four- or six-cylinder units provided a range of six models. A mechanical feature of great interest was the built-up crankshaft, of which each overhung throw was a separate forging bolted to its neighbour—a construction that made possible the use of large-diameter roller bearings between the webs. Less elaborate and more conventional crankshafts were used later however.

As part of its toroidal-chamber combustion system, Saurer also introduced its own design of four-hole injectors, so constructed that they could not be incorrectly assembled again after dismantling. Each injector is carried in a housing screwed into the cylinder head and passing through the water jacket. The needle is formed with two diameters, the smaller part having a clearance within the nozzle through which fuel can flow to the tip, while the larger is a close sliding fit in the guide within the nozzle body and is well removed from the hottest zone. Four spray holes of 0·25 mm (0·009 in) bore are spaced around the nozzle tip. This type of injector and the method of mounting it shows an early realisation of the cooling requirements of this component, and this example has since been widely followed.

Experiments with exhaust-driven turbochargers began in 1937 and resulted in a power increase of 35 to 45%. A six-cylinder 8 l. (488 in^3) engine developing 100 bhp at 1 900 rev/min when naturally aspirated was stepped up to 135 bhp at the same speed while retaining the very favourable specific fuel consumption of 0·37 lb/bhp h (167 g/bhp h).

For the heaviest classes of road vehicles, V8 engines of 110 mm (4·33 in) bore and 140 mm (5·51 in) stroke were produced before 1939, while a 1947 show saw the introduction of a 60 degree V12 engine of 15·91 l. (970 in^3) capacity and 300 bhp output, having a Brown-Boveri turbocharger for each bank of cylinders. An unusual feature of this unit was that the inlet valves could be rotated by an external control so that the position of the masks under each valve head could be adjusted to modify the swirl of the ingoing air in relation to load and fuel delivery characteristics.

Some years later, the company introduced a four-cylinder engine of 115 mm (4·53 in) bore and 140 mm (5·51 in) stroke, having a capacity of 5·82 l. (366 in^3) which was boosted by a belt-driven supercharger of the screw-lobe Lysholm-Smith type. One rotor of the supercharger was lobed and the other grooved, both lobes and grooves being helical and meshing together as they rotated in opposite directions. This engine developed 110 bhp at 1 900 rev/min. A derivative of this engine was a V8 unit of 11·63 l. (710 in^3), embodying two of the four-cylinder banks and similarly supercharged, which was rated at 200 bhp at 2 000 rev/min.

In the mid-1950s, Saurer introduced its first underfloor engines, derived from vertical units then in production. These were not, however, the first Swiss engines of horizontal type. One model of 10·3 l. (628 in^3) capacity, with six cylinders of 125 mm bore and 140 mm stroke (4·92 in by 5·51 in), marked the reappearance of four valves per cylinder, although a two-valve version was also made. This arrangement of a four-valve head and screw-lobe supercharger is still in production today.

All Saurer automotive engines in current production are of six-cylinder (4·53 in) giving a capacity of 7·48 l. (416 in^3). It develops 130 bhp (DIN) at 2 200 rev/min and has a maximum torque of 46 kgf m (333 lbf ft). The 140 mm stroke engine also has a 115 mm (4·53 in) bore, its capacity being 8·725 l. (532 in^3), its power 160 bhp at 2 200 rev/min and its torque 60 kgf m (434 lbf ft).

The 150 mm stroke engine is made either normally aspirated, with a 130 mm (5·12 in) bore—giving 11·95 l. (730 in^3) and a power of 230 bhp—or mechanically supercharged in a 128 mm (5·04 in) bore version; this 11·6 l. (706 in^3) engine develops 275 bhp at 2 000 rev/min and its maximum torque is 108 kgf m (783 lbf ft). Turbocharged, it produces 310 bhp at 2 200 rev/min, a torque of 114 kgf m (825 lbf ft) and the good fuel consumption of 157 g/bhp h (0·346 lb/bhp h).

SCANIA (AB Scania-Vabis, Södertälje, Sweden)

This old-established Swedish engineering and vehicle-building firm began to make industrial diesel engines in 1936 and produced its first direct-injection engine in 1950. There are now three basic in-line engines in the Scania range and both a normally aspirated and turbocharged version of each. These have now been joined by a turbocharged V8.

Two of the basic models, the D5 four-cylinder (5·2 l.) and the D8 six-cylinder (7·79 l.) share the same bore and stroke of 115 mm (4·53 in) and 125 mm (4·92 in). In normally aspirated form the four-cylinder develops 102 bhp

The turbocharged version of the 7·79 l. (475 in^3) Scania D8

(DIN) at 2 400 rev/min, but when turbocharged the power is 130 bhp. The normally aspirated six-cylinder develops 145 bhp at 2 400 rev/min while in turbocharged form the power is 192 bhp.

The biggest in-line D11, has a bore and a stroke of 127 mm by 145 mm (5 in × 5·71 in), there being six-cylinders in-line, giving a capacity of 11·02 l.

(674 in^3). In normally aspirated form the output is 195 bhp (DIN) at 2 200 rev/min. When turbocharged this rises to 275 bhp which is a substantial boost. In achieving the extra power the minimum specific fuel consumption improves from 0·379 to 0·355 lb/bhp h (171 to 161 g/bhp h).

Induction and exhaust ports are on opposite sides of the heads of Scania engines. When a Holset turbocharger is fitted it is mounted on top of the centre of the exhaust manifold. The fuel injectors are outside the rocker covers. Wet cylinder liners on all the Scania engines are claimed to give slightly lower piston temperatures, which can be critical with turbocharged engines. There is a separate oil feed, via a filter, to the turbocharger from the engine's lubrication system. Auxiliaries are gear driven from the front-end timing device,

The exhaust brake control on the turbocharged DS11 Scania engine is linked to the fuel pump 'stop' lever

except for twin belt drive to the fan and alternator. A Glacier centrifugal oil filter and a water–oil heat exchanger are included in the lubrication system.

The pistons have a pronounced wide-section cone forming the middle of the toroidal cavity. Three compression and one scraper ring are fitted to each piston, the top ring being chromium plated and sitting in an Alfin cast-iron insert. Steel gaskets are used on the big engines but the smaller sizes have copper/asbestos gaskets.

Fuel injection pumps on turbocharged engines have a diaphragm on the end of the rack; being piped to the inlet manifold and therefore sensitive to the induction pressure, it alters the maximum pump-rack travel according to the gas speed and prevents excess fuel being delivered to the fuel injectors until there is enough air passing to the combustion chambers to burn the fuel.

The turbocharged V8 is called the DS14. The maximum power from its 14·2 l. (868 in^3) is 350 bhp (DIN) at 2 300 rev/min, while the maximum torque is 957 lbf ft (127 kgf m) at 1 450 rev/min. Like the in-line Scanias, it has wet cylinder liners, toroidal piston cavities and helical timing gears. There is a separate head for each cylinder.

STEYR (Steyr-Daimler-Puch A.G., Vienna, Austria)

This Austrian chassis manufacturer broke with tradition in 1967 when it brought out its largest engine, an 8·14 l. (496 in^3) in-line six-cylinder unit designated the WD 614. Previous Steyr engines had all been of antechamber design, incorporating a 'pepper-pot' combustion chamber in the cylinder head, but this unit appeared with toroidal-cavity combustion chambers in the piston crowns.

To increase power output and yet keep the maximum piston speed within reasonable bounds, the bore dimension was made equal to the piston stroke, both being 4·72 in (120 mm). At the same time, the engine was designed to resist the extra mechanical and thermal stresses imposed by turbocharging and, in fact, the first turbocharged version appeared at the Frankfurt Show of the same year. This raised the DIN power from 180 to 230 bhp (at 2 600

Steyr WD 614 direct-injection engine, incorporating individual cylinder heads and thermostatically controlled fan

instead of 2 800 rev/min). Other design features include specially shaped inlet ducts in the individual cylinder heads, to impart a high degree of air swirl, replaceable dry cylinder liners and a fan incorporating a thermostatically-controlled coupling. The minimum specific fuel consumption is 170 g/bhp h (0·374 lb/bhp h).

A smaller six-cylinder, the WD 610, has a bore of 105 mm (4·13 in) and a stroke of 115 mm (4·53 in) and is a direct-injection development of an earlier indirect-injection design. It is available either normally aspirated, governed at 2 800 rev/min with a power output of 132 bhp (DIN) or turbocharged when it develops 150 bhp at 2 600 rev/min. This engine has three cylinder heads, each covering two bores.

The latest engine is a V8, available in either naturally aspirated or turbocharged forms (when a separate turbocharger serves each bank of cylinders). It has a singularly good power-to-size ratio, obtaining 320 bhp (DIN) at

2 600 rev/min from 12 l. with a maximum torque of 100 kgf m (723 lbf ft). In naturally aspirated form the power is 260 bhp with a torque of 81 kgf m, or 586 lbf ft. This V8, the WD 815, is a development of the six-cylinder WD 614 series, with similar basic features such as the individual cylinder heads, dry liners and equalised bore and stroke.

TATRA (Národní Podnik Koprivnice, Nositel Rádu, Prague, Czechoslovakia)

Air-cooled diesel engines are the speciality of Tatra and three new designs have been introduced. The two principal automotive units are the T928 and T930, both used in industrial and railway applications. They are both 75 degree vee-form engines, but the T928 is an 11·76 l. (718 in^3) V8 and the T930 is a 17·64 l. V12. Neither speeds nor powers are high. The V8 develops 192 bhp (DIN) at 2 000 rev/min and the V12, 250 bhp at the same speed. A bore of 120 mm (4·72 in) and stroke of 130 mm (5·12 in) are common to both engines.

A Tatra V12 air-cooled diesel

Direct-injection gives fair fuel consumption, the V8's minimum specific fuel consumption is 0·385 and the V12's is 0·374 lb/bhp h (173 and 169 g/bhp h).

For their capacities, these Tatra air-cooled engines are comparatively light. The T928 weighs only 1 430 lb and the T930, 2 280 lb (648 and 1034 kg). Aluminium-alloy cylinder heads sit on cast-iron blocks without a gasket between them; unusually the crankshaft revolves in roller bearings. The timing drive is by gears at the front of the engine, while a cowled, multi-bladed fan blows air between the vee of the cylinder banks, from where it escapes to the outside of the vee past the finned heads and cylinder barrels.

The T924 air-cooled engine is a four-cylinder in-line design but its general construction follows that of the vee-form engines and it has the same bore and stroke. It develops 70 bhp at 1 800 rev/min from its 5·87 l. (358 in^3).

Despite the name Tatra being traditionally associated with air-cooled engines, a water-cooled V8, again with a 75 degree vee angle, has now been introduced. It has the same 120 mm (4·72 in) bore of the air-cooled units, but a longer stroke—140 mm (5·51 in). This gives a capacity of 12·66 l. (772 in^3) producing a maximum power output of 200 bhp at 2 000 rev/min and a maximum torque of 593 lbf ft at 1 200 rev/min.

Apart from being water-cooled this V8, called the T2–138, shares many of the design features of the air-cooled engines, such as the roller-bearing mains, aluminium heads and cast-iron block with no gaskets.

UNIC [Simca Industries SA, 1 Quai National, Puteaux (Seine) 506, France (part of Fiat)]

In the bigger Unic engines one is fairly unusual in its cylinder arrangement. The in-line engine has five cylinders and the latest engine on which a big sales drive is being mounted is a V8. Both are four-stroke, direct-injection engines with toroidal piston cavities. The five-cylinder engine develops 140 bhp at 2 600 rev/min. Its capacity is 6·73 l. (410 in^3), the bore being 119 mm (4·69 in) and the stroke 121 mm (4·76 in). It has a silicone-fluid vibration damper on the crankshaft. The minimum specific fuel consumption is 165 g/bhp h (0·364 lb/bhp h).

When first introduced the V8 engine also had a bore of 119 mm and a stroke of 121 mm. In this case the capacity was 10·766 l. (657 in^3) and the maximum power on the SAE rating 270 bhp. Now, a 14·88 l. (910 in^3) edition has been introduced, with a bore of 135 mm (5·32 in) and a stroke of 130 mm (5·12 in). This has an SAE power rating of 340 bhp (310 bhp, DIN) and a torque of

The V8 engine in a Unic tilt-cab truck

110 kgf m (796 lbf ft). The governed speed is 2 400 rev/min and fuel consumption is 162 g/bhp h (0·356 lb/bhp h). The engine is installed in Italian OM as well as Unic heavy trucks. Both Unic and OM are controlled by Fiat.

This 90 degree V8 design has a cast-iron block and heads, there being two heads for each bank. The inlet manifolds are on the inside of the V and the exhaust manifolds on the outside. The fuel injectors are also on the inside of the vee. There are two duplex belt drives, one of which drives the alternator and air compressor while the other one drives the fan and water pump. A Holset viscous vibration damper is fitted to the front of the crankshaft. The oil-circulation system includes two full-flow filters. The 14·88 l. (910 in^3) V8 weighs 990 kg (2 180 lb).

The two, rather more conventional units are in-line four-cylinder designs. They both have a bore of 119 mm but the stroke is different. The 5·38 l. model has a stroke of 121 mm while the 4·493 l. (469 in^3) version is oversquare, having a stroke of 101 mm (3·98 in). Both engines have a compression ratio of 17:1 and use the Saurer direct-injection combustion system. They have two cylinder heads and induction-hardened crankshafts with five copper/lead bearings.

The 5·38 l. (328 in^3) engine develops 135 bhp at 2 600 rev/min on the SAE rating and a maximum torque of 42 kgf m (304 lbf ft) at 1 400 rev/min.

VOLVO (AB Volvo, Gothenburg, Sweden)

Turbocharging figures strongly in the Volvo engine programme, for all four basic engines have been designed with turbocharging in mind. The latest unit, in fact, made its debut in turbocharged form only. A thriving business has been built on marine equivalents of these engines as well. All Volvo engines are of in-line six-cylinder configuration, employing direct-injection in toroidal piston cavities. They take a conventional form, with cast-iron cylinder blocks and heads, front-end timing gears, induction-hardened crankshafts, wet cylinder liners and two valves per cylinder.

The present generation of engines was introduced in 1965. It comprised the D50A, of 5·1 l. (311 in^3), the 6·7 l. (409 in^3) D70A and the 9·6 l. (586 in^3) D100A. Turbocharged versions were given the prefix letter 'T' when they were brought in later. With a bore of 95·25 mm (3·75 in) and stroke of 120 mm (4·72 in), the 5·1 l. engine gives 107 bhp when normally aspirated and 147 bhp when turbocharged. Maximum torque is, respectively, 249 lbf ft at 1 600 rev/min or 318 lbf ft (43·9 kgf m) at 1 700 rev/min.

There is no common bore or stroke dimension among the three basic engines, but the stroke lengthens in 10 mm (0·39 in) increments, being 120, 130 and 140 mm (4·72, 5·12 and 5·51 in). Bore dimensions are 95·3, 104·8 and 120·7 mm (3·75, 4·12 and 4·75 in). In naturally aspirated form, the three engines develop 107 bhp at 2 800 rev/min, 144 bhp at 2 500 rev/min and 166 bhp at 2 200 rev/min. Addition of a turbocharger boosts these outputs to 147, 195 (at 2 400 rev/min) and 260 bhp, which represent increases of 37·4, 35·4 and no less than 56·6%. The 9·6 l. (586 in^3) engine is also produced in a horizontal version for use in buses and coaches.

Lately, Volvo has strengthened the 6·7 l. (409 in^3) engine, and made modifications to increase power output and restrict smoke emission when the

The 192 bhp turbocharged Volvo TD70B

engine is pulling hard. The modifications include a new camshaft, a new design of piston, and a new governor on the fuel-injection pump. In this improved form the engine becomes the D70B, or TD70B, when turbocharged. Respective power outputs are 150 bhp at 2 500 rev/min and 192 bhp at 2 400 rev/min.

Volvo's latest introduction is the TD120 engine, which is a turbocharged 12 l. (732 in^3) unit of 130·2 mm (5·125 in) bore and 150 mm (5·91 in) stroke intended for powering its heaviest range of trucks designed for operation at 50 tons gross weight. Like the 9·6 l. engine, this has separate heads for each cylinder, each being held down by eight bolts spaced evenly around the bore. Although a high output of 330 bhp is developed at 2 200 rev/min, accompanied by a maximum torque of 927 lbf ft (128 kgf m), the turbocharging system is quite straightforward, without charge cooling. Considerable attention, however, has been given to a high volume flow of coolant and lubricating oil to avoid any local overheating at critical points. The dry weight of this engine is 2 284 lb (1 036 kg).

YMZ (Yaroslavl Motor Plant, Russia—via Avtoexport, 32 Smolenskaja Pl., Moscow G-200)

Tremendous strides have been made in Russian automobile engineering techniques over the past seven years or more. Simple testimony to this is provided by the two diesel engines on which the big YMZ concern is now concentrating. A remarkable degree of rationalisation has also been attained between the two engine models, which are direct-injection vee-form types. It is claimed that 93% of the parts are common to both models.

Both YMZ engines have a bore of 130 mm (5·12 in) and a stroke of 140 mm (5·5 in) and are 90 degree vees. The YMZ 236 is a six-cylinder 11·15 l. (680 in^3) engine developing 180 bhp (DIN) at 2 100 rev/min. The 238 model is an eight-cylinder 14·86 l. (907 in^3) unit developing 240 bhp at 2 100 rev/min and weighing a ton. Inlet and exhaust valves are on opposite sides of each single-piece cast-iron head and there are two valves per cylinder. The valves are push-rod operated via roller-ended rocker-tappets on a single camshaft housed in a tunnel running down the intersection of the vee. An in-line fuel injection pump between the two banks of cylinders delivers fuel to injectors mounted vertically beneath the rocker covers.

Steel/asbestos gaskets seal the joint between head and cast-iron block. Wet cylinder liners are employed, and the pistons have fairly shallow toroidal piston cavities. Each veed pair of connecting rods shares a common crankpin,

The YMZ-238 V8 engine with synchromesh gearbox

side by side, while the big ends are split at an angle so that they can be withdrawn through the bore. The helical-gear timing drive at the front of these engines is unusual in having tandem sets of gears, the back set taking the drive from camshaft to fuel injecuion pump and the front set a drive to the large water pump and fan. Vee belts are not eliminated, however, as both the fan shaft and the crankshaft-nose carry pulleys. Two short belts from the fan shaft drive the generator and the air compressor; another short belt from the crankshaft pulley driving the power steering pump.

A lower gear drive from the timing gear on the crankshaft drives a double oil pump, circulating oil through a cooler as well as through the engine's lubrication galleries. Turbocharged versions of these V8 designs are being developed. The turbocharger is mounted transversely between the cylinder banks at the rear and is reputed to boost the power of the V8 to 300 bhp.

Appendices

MAKE	*MODEL*	*COOLING* *water or air-cooled*	*CYCLE* *4-stroke or 2-stroke*	*ASPIRATION* *normally asp. super-charged or turbo-charged*	*NUMBER OF CYLINDERS*	*LAYOUT* *in-line, vee or horizontally opposed*	*BORE* *in (mm*
AEC	A505	water	4	N.A.	6	in-line	4·50 (114)
AEC	A691	water	4	N.A.	6	in-line	5·12 (130)
AEC	A760	water	4	N.A.	6	in-line	5·37 (136)
AEC	AV1100	water	4	N.A.	6	in-line	6·13 (156)
AEC	AVT1100	water	4	Turb.	6	in-line	6·13 (156)
AEC	801 Vee	water	4	N.A.	8	90° Vee	5·31 (135)
BEDFORD	70	water	4	N.A.	6	in-line	4·56 (116)
BEDFORD	220	water	4	N.A.	4	in-line	4·06 (103)
BEDFORD	330	water	4	N.A.	6	in-line	4·06 (103)
BEDFORD	60	water	4	N.A.	6	in-line	4·13 (105)
BERLIET	M420–30	water	4	N.A.	4	in-line	4·72 (120)
BERLIET	M420	water	4	N.A.	4	in-line	4·72 (120)
BERLIET	M520	water	4	N.A.	5	in-line	4·72 (120)
BERLIET	M620	water	4	N.A.	6	in-line	4·72 (120)
BERLIET	M635–CD	water	4	N.A.	6	in-line	5·31 (135)
BERLIET	M640	water	4	N.A.	6	in-line	5·51 (140)

TROKE n (mm)	CAPACITY in^3 (l.)	MAX. BHP BS (or DIN) at what rev/min	MAX. BHP SAE at what rev/min	MAX. TORQUE BS (or DIN) at what rev/min lbf ft (kgf m)	TORQUE AT GOV- ERNED SPEED BS (or DIN) lbf ft (kgf m)	MINIMUM SPECIFIC FUEL CONSUMP- TION at full load lb/bhp h (g/bhp h)	DRY WEIGHT with flywheel and starter, but no fan, generator, air filter or clutch lb (kg)	MODEL
5·12 (130)	502 (8·19)	151 2 400	162 —	384 (53) 1 300	329 (45·4)	0·36 (163)	1 365 (620)	A505
5·59 (142)	690 (11·31)	200 2 200	215 —	558 (77·3) 1 200	477 (66·0)	0·36 (163)	1 890 (858)	A691
5·59 (142)	761 (12·47)	220 2 200	236 —	600 (83·1) 1 400	525 (72·7)	0·36 (163)	1 904 (865)	A760
6·13 (156)	1 083 (17·75)	289 1 900	310 —	930 (128·9) 1 300	800 (110·8)	0·37 (168)	3 375 (1 529)	AV1100
6·13 (156)	1 083 (17·75)	377 1 900	404 —	1 230 (170·4) 1 200	1 041 (144·2)	0·37 (168)	3 475 (1 576)	AVT1100
4·50 (115)	799 (13·10)	272 2 600	291 —	638 (88·1) 1 400	549 (76·0)	0·37 (168)	2 050 (930)	801 Vee
4·75 (121)	466 (7·63)	138 2 500	146 —	321 (44·5) 1 000	304 (42·3)	0·36 (163)	1 395 (634)	70
4·25 (108)	220 (3·61)	64 2 800	70 —	153 (21·2) 1 400	131 (18·1)	0·39 (177)	775 (352)	220
4·25 (108)	330 (5·42)	107 2 800	112 —	234 (32·4) 1 800	200 (27·7)	0·38 (172)	998 (453)	330
4·75 (120)	381 (6·24)	115 2 800	124 —	258 (35·8) 1 400	218 (30·2)	0·37 (168)	1 340 (608)	60
5·12 (130)	359 (5·88)	120 2 600	128 2 600	268 (37) 1 500	217 (30)	0·37 (168)	1 147 (520)	M420–30
5·51 (140)	384 (6·30)	—	120 2 100	296 (41) 1 400	271 (37·5)	0·37 (168)	1 608 (730)	M420
5·51 (140)	482 (7·90)	—	150 2 100	376 (52) 1 400	340 (47)	0·37 (168)	1 915 (870)	M520
5·51 (140)	580 (9·50)	—	180 2 100	448 (62) 1 400	415 (57·5)	0·37 (168)	2 200 (1 000)	M620
5·51 (140)	732 (12·02)	—	250 2 200	658 (91) 1 300	560 (77·3)	0·37 (168)	2 200 (1 000)	M635–CD
6·30 (160)	903 (14·80)	—	240 1 800	686 (95) 1 300	635 (88)	0·39 (177)	2 890 (1 310)	M640

MAKE	*MODEL*	*COOLING* *water or air-cooled*	*CYCLE* *4-stroke or 2-stroke*	*ASPIRATION* *normally asp. super-charged or turbo-charged*	*NUMBER OF CYLINDERS*	*LAYOUT* *in-line, vee or horizontally opposed*	*BORE* *in (mm*
BERLIET	M640	water	4	Turb.	6	in-line	5·51 (140)
BERLIET	MIS645–50	water	4	Turb.	6	in-line	5·71 (145)
BERLIET	MS840	water	4	Turb.	8	90° Vee	5·51 (140)
BERLIET	V825	water	4	N.A.	8	90° Vee	4·92 (125)
BERLIET	V800	water	4	N.A.	8	90° Vee	3·94 (100)
BÜSSING	S7D	water	4	N.A.	6	in-line	4·25 (108)
BÜSSING	U7D	water	4	N.A.	6	in-line	4·25 (108)
BÜSSING	S11D	water	4	N.A.	6	in-line	5·04 (128)
BÜSSING	U11D	water	4	N.A.	6	in-line	5·04 (128)
BÜSSING	U12D	water	4	N.A.	6	in-line	5·20 (132)
BÜSSING	U12DA	water	4	Turb.	6	in-line	5·20 (132)
CATERPILLAR	1673CT	water	4	Turb.	6	in-line	4·75 (121)
CATERPILLAR	1674TA	water	4	Turb. intercool	6	in-line	4·75 (121)
CATERPILLAR	1693T	water	4	Turb.	6	in-line	5·40 (137)

TROKE n (mm)	CAPACITY in^3 (l.)	MAX. BHP BS (or DIN) at what rev/min	MAX. BHP SAE at what rev/min	MAX. TORQUE BS (or DIN) at what rev/min lbf ft (kgf m)	TORQUE AT GOV- ERNED SPEED BS (or DIN) lbf ft (kgf m)	MINIMUM SPECIFIC FUEL CONSUMP- TION at full load lb/bhp h (g/bhp h)	DRY WEIGHT with flywheel and starter, but no fan, generator, air filter or clutch lb (kg)	MODEL
6·30 (160)	903 (14·80)	—	315 1 800	854 (118) 1 400	795 (110)	0·38 (172)	2 980 (1 350)	M640
5·91 (150)	906 (14·86)	—	400 2 300	998 (138) 1 300	904 (125)	0·39 (177)	3 300 (1 500)	MIS645–50
6·30 (160)	1 200 (19·70)	—	420 1 800	1 208 (167) 1 400	1 140 (158)	0·38 (172)	3 640 (1 650)	MS840
5·12 (130)	779 (12·76)	300 2 500	—	695 (96) 1 600	660 —	0·37 (168)	2 307 (1 050)	V825
4·34 (110)	421 (6·92)	170 3 000	—	355 (49) 1 700	312 (43)	0·38 (172)	—	V800
5·32 (135)	453 (7·42)	150 2 400	162 —	347 (48) 1 400	318 (44)	0·36 (163)	1 370 (620)	S7D
5·32 (135)	453 (7·42)	156 2 400	167 —	354 (49) 1 400	332 (46)	0·36 (163)	1 410 (640)	U7D
		135 2 400	142 —	318 (44) 1 400	295 (41)	0·35 (159)	1 410 (640)	
5·91 (150)	707 (11·58)	210 2 100	230 —	614 (84) 1 300	520 (72)	0·36 (163)	2 095 (950)	S11D
5·91 (150)	707 (11·58)	210 2 100	230 —	614 (84) 1 300	520 (72)	0·36 (163)	2 070 (940)	U11D
		217 2 200	235 —	577 (80) 1 300	518 (71·5)	0·37 (168)	2 070 (940)	
5·91 (150)	752 (12·32)	240 2 200	260 —	621 (86) 1 300	563 (78)	0·36 (163)	2 160 (980)	U12D
5·91 (150)	752 (12·32)	280 2 200	305 —	723 (100) 1 300	636 (88)	0·34 (155)	2 300 (1 045)	U12DA
		310 2 200	340 —	853 (118) 1 300	723 (100)	0·35 (159)	2 300 (1 045)	
6·00 (152)	641 (10·50)	225 —	250 2 200	690 (94) 1 550	569 (78·6)	0·40 (182)	1 940 (880)	1673CT
6·00 (152)	641 (10·50)	225 —	270 2 200	745 (103) 1 400	605 (83·5)	0·39 (177)	2 260 (1 025)	1674TA
6·50 (165)	893 (14·60)	—	325 2 100	1 000 (139) 1 485	778 (108)	0·38 (172)	3 225 (1 470)	1693T

MAKE	*MODEL*	*COOLING water or air-cooled*	*CYCLE 4-stroke or 2-stroke*	*ASPIRATION normally asp. super-charged or turbo-charged*	*NUMBER OF CYLINDERS*	*LAYOUT in-line, vee or horizontally opposed*	*BORE in (mm*
CATERPILLAR	1693TA	water	4	intercool	6	in-line	5·40 (137)
CATERPILLAR	1145	water	4	N.A.	8	90° Vee	4·50 (114)
CATERPILLAR	1150	water	4	N.A.	8	90° Vee	4·50 (114)
CATERPILLAR	1160	water	4	N.A.	8	90° Vee	4·50 (114)
CHRYSLER U.K. (Ex ROOTES)	3D-215	water	2	Scav Sup.	3	in-line opp. piston	3·38 (86)
CUMMINS	V6-155	water	4	N.A.	6	90° Vee	4·63 (117)
CUMMINS	V8-210	water	4	N.A.	8	90° Vee	4·63 (117)
CUMMINS	V8-555	water	4	N.A.	8	90° Vee	4·63 (117
CUMMINS	V6-200	water	4	N.A.	6	90° Vee	5·50 (140)
CUMMINS	NH-220	water	4	N.A.	6	in-line	5·13 (130
CUMMINS	NTO-6	water	4	Turb.	6	in-line	5·13 (130
CUMMINS	NRTO-6	water	4	Turb.	6	in-line	5·13 (130
CUMMINS	NHRS-6	water	4	Sup.	6	in-line	5·13 (130
CUMMINS	NHS-6	water	4	Sup.	6	in-line	5·13 (130
CUMMINS	V8-265	water	4	N.A.	8	90° Vee	5·50 (140
CUMMINS	NH-250	water	4	N.A.	6	in-line	5·50 (140
CUMMINS	NT-280	water	4	Turb.	6	in-line	5·50 (140

STROKE in (mm)	*CAPACITY in³ (l.)*	*MAX. BHP BS (or DIN) at what rev/min*	*MAX. BHP SAE at what rev/min*	*MAX. TORQUE BS (or DIN) at what rev/min lbf ft (kgf m)*	*TORQUE AT GOV-ERNED SPEED BS (or DIN) lbf ft (kgf m)*	*MINIMUM SPECIFIC FUEL CONSUMP-TION at full load lb/bhp h (g/bhp h)*	*DRY WEIGHT with flywheel and starter, but no fan, generator, air filter or clutch lb (kg)*	*MODEL*
6·50 (165)	893 (14·60)	—	375 2 000	1 145 (159) 1 475	936 (129)	0·37 (168)	3 335 (1 513)	1693TA
4·10 (104)	522 (8·56)	162 3 200	175 —	352 (49) 1 700	279 (38·6)	0·38 (172)	1 196 (544)	1145
4·50 (114)	573 (9·40)	188 3 000	200 —	446 (61·5) 1 400	344 (47·6)	0·37 (168)	1 196 (544)	1150
5·00 (127)	636 (10·40)	208 2 800	225 —	510 (70·5) 1 400	409 (565)	0·37 (168)	1 196 (544)	1160
4×2 (102 ×2)	215 (3·52)	125 2 300	145 —	335 (46·5) 1 200	298 (41)	0·38 (172)	1 075 (488)	3D-215
3·75 (95)	378 (6·20)	155 3 300	—	302 (41·7) 1 900	258 (35·7)	0·39 (177)	1 196 (544)	V6-155
3·75 (95)	504 (8·27)	210 3 300	—	405 (56) 1 900	350 (48·4)	0·39 (177)	1 460 (665)	V8-210
4·125 (105)	555 (9·1)	240 3 300	—	445 (61·8) 1 900	383 (53)	0·38 (172)	1 700 (782)	V8-555
4·13 (105)	588 (9·64)	200 2 600	—	451 (62·5) 1 800	405 (56·1)	0·37 (168)	1 690 (767)	V6-200
6·00 (152)	743 (12·17)	205 or 220 2 100	—	612 (84·8) 1 400	550 (76·2)	0·40 (182)	2 360 (1 070)	NH-220
6·00 (152)	743 (12·17)	262 2 100	—	720 (99·6) 1 400	655 (90·7)	0·38 (172)	2 455 (1 112)	NTO-6
6·00 (152)	743 (12·17)	355 2 100	—	900 (125) 1 500	837 (116)	0·41 (186)	2 455 (1 112)	NRTO-6
6·00 (152)	743 (12·17)	320 2 100	—	860 (119) 1 600	800 (111)	0·44 (200)	2 660 (1 209)	NHRS-6
6·00 (152)	743 (12·17)	290 2 100	—	775 (107) 1 550	725 (100)	0·45 (204)	2 660 (1 208)	NHS-6
4·13 (105)	785 (12·85)	265 2 600	—	600 (83·0) 1 800	535 (74·1)	0·37 (168)	2 080 (945)	V8-265
6·00 (152)	855 (14·00)	240 2 100	250 —	620 (94·2) 1 400	625 (86·5)	0·40 (182)	2 490 (1 150)	NH-250
6·00 (152)	855 (14·00)	270 2 100	280 —	750 (108) 1 500	700 (96·9)	0·38 (172)	2 650 (1 200)	NT-280

MAKE	MODEL	COOLING water or air-cooled	CYCLE 4-stroke or 2-stroke	ASPIRATION normally asp. supercharged or turbocharged	NUMBER OF CYLINDERS	LAYOUT in-line, vee or horizontally opposed	BORE in (mm)
CUMMINS	NT-310	water	4	Turb.	6	in-line	5·50 (140)
CUMMINS	NT-335	water	4	Turb.	6	in-line	5·50 (140)
CUMMINS	NT-380	water	4	Turb.	6	in-line	5·50 (140)
DAF	DA475	water	4	N.A.	6	in-line	3·98 (100)
DAF	DD575	water	4	N.A.	6	in-line	3·98 (100)
DAF	DS575	water	4	Turb.	6	in-line	3·98 (100)
DAF	DF615	water	4	N.A.	6	in-line	4·09 (104)
DAF	DH825	water	4	N.A.	6	in-line	4·70 (118)
DAF	DU825	water	4	Turb.	6	in-line	4·70 (118)
DAF	DKA1160	water	4	N.A.	6	in-line	5·12 (130)
DAF	DKD1160	water	4	N.A.	6	in-line	5·12 (130)
DAF	DK1160	water	4	N.A.	6	in-line	5·12 (130)
DAF	DKB1160	water	4	Turb.	6	in-line	5·12 (130)
DAIMLER-BENZ	OM615	water	4	N.A.	4	in-line	4·42 (111)
DAIMLER-BENZ	OM314	water	4	N.A.	4	in-line	3·82 (97)
DAIMLER-BENZ	OM352	water	4	N.A.	6	in-line	3·82 (97)
DAIMLER-BENZ	OM352A	water	4	Turb.	6	in-line	3·82 (97)

STROKE in (mm)	*CAPACITY in³ (l.)*	*MAX. BHP BS (or DIN) at what rev/min*	*MAX. BHP SAE at what rev/min*	*MAX. TORQUE BS (or DIN) at what rev/min lbf ft (kgf m)*	*TORQUE AT GOV-ERNED SPEED BS (or DIN) lbf ft (kgf m)*	*MINIMUM SPECIFIC FUEL CONSUMP-TION at full load lb/bhp h (g/bhp h)*	*DRY WEIGHT with flywheel and starter, but no fan, generator, air filter or clutch lb (kg)*	*MODEL*
6·00 (152)	855 (14·00)	310 2 100	322 —	850 (118) 1 500	775 (107)	0·38 (172)	2 750 (1 247)	NT-310
6·00 (152)	855 (14·00)	335 2 100	349 —	930 (128) 1 600	837 (116)	0·38 (172)	2 750 (1 247)	NT-335
6·00 (152)	855 (14·00)	364 2 300	380 —	955 (132) 1 650	870 (120)	0·40 (182)	2 750 (1 247)	NT-380
3·98 (100)	291 (4·77)	90 2 500	100 —	203 (27·5) 1 600	174 (24)	0·41 (185)	916 (415)	DA475
4·76 (120)	351 (5·75)	110 2 400	120 —	256 (35·6) 1 400	238 (33)	0·39 (177)	1 136 (515)	DD575
4·76 (120)	351 (5·75)	150 2 400	165 —	354 (49) 1 000	325 (44)	0·37 (168)	1 190 (540)	DS575
4·76 (120)	415 (6·17)	126 2 600	138 —	275 (38) 1 400	253 (35)	0·38 (170)	1 158 (525)	DF615
4·96 (126)	503 (8·25)	156 2 400	172 —	390 (54) 1 400	340 (47)	0·36 (164)	1 368 (620)	DH825
4·96 (126)	503 (8·25)	201 2 400	218 —	477 (66) 1 600	— (60)	0·36 (163)	1 440 (655)	DU825
5·75 (146)	708 (11·60)	230 2 200	250 —	593 (82) 1 300	542 (75)	0·37 (168)	1 810 (820)	DKA1160
5·75 (146)	708 (11·60)	165 2 000	180 —	484 (67) 1 000	434 (60)	0·35 (159)	1 820 (820)	DKD1160
5·75 (146)	708 (11·60)	212 2 200	230 —	527 (73) 1 200	498 (69)	0·36 (163)	1 810 (820)	DK1160
5·75 (146)	708 (11·60)	304 2 200	324 —	813 (105) 1 500	760 (101)	0·35 (158)	1 878 (850)	DKB1160
3·64 (92)	134 (2·20)	60 4 200	64 —	83 (11·5) 2 400	66 (9·1)	0·42 (190)	405 (184)	OM615
5·04 (128)	231 (3·78)	85 2 800	90 —	161 (23) 1 600	148 (20·5)	0·35 (159)	660 (300)	OM314
5·04 (128)	346 (5·68)	130 2 800	140 —	260 (36) 1 600	233 (32·2)	0·35 (159)	900 (410)	OM352
5·04 (128)	346 (5·68)	168 2 800	185 —	346 (48) 1 600	316 (43·8)	0·35 (158)	930 (422)	OM352A

MAKE	*MODEL*	*COOLING* *water or air-cooled*	*CYCLE* *4-stroke or 2-stroke*	*ASPIRATION* *normally asp. super-charged or turbo-charged*	*NUMBER OF CYLINDERS*	*LAYOUT* *in-line, vee or horizontally opposed*	*BORE* *in (mm)*
DAIMLER-BENZ	OM346	water	4	N.A.	6	in-line	5·04 (128)
DAIMLER-BENZ	OM346	water	4	Sup.	6	in-line	5·04 (128)
DAIMLER-BENZ	OM355	water	4	N.A.	6	in-line	5·04 (128)
DAIMLER-BENZ	OM360	water	4	N.A.	6	in-line	4·53 (115)
DAIMLER-BENZ	OM403	water	4	N.A.	10	90° Vee	4·92 (125)
DAIMLER-BENZ	OM402	water	4	N.A.	8	90° Vee	4·92 (125)
DEUTZ	F4L912	air	4	N.A.	4	in-line	3·94 (100)
DEUTZ	F6L912	air	4	N.A.	6	in-line	3·94 (100)
DEUTZ	F6L413	air	4	N.A.	6	90° Vee	4·73 (120)
DEUTZ	F8L413	air	4	N.A.	8	90° Vee	4·72 (120)
DEUTZ	F10L413	air	4	N.A.	10	90° Vee	4·72 (120)
DEUTZ	F12L413	air	4	N.A.	12	90° Vee	4·72 (120)
DORMAN	8JV	water	4	N.A.	8	90° Vee	5·12 (130)
F.B.W.	CUA	water	4	Turb.	6	in-line	4·17 (106)
F.B.W.	E	water	4	N.A.	6	in-line	4·92 (125)
F.B.W.	E3	water	4	N.A.	6	in-line	5·04 (128)
F.B.W.	ET	water	4	Turb.	6	in-line	4·92 (125

STROKE in (mm)	*CAPACITY in³ (l.)*	*MAX. BHP BS (or DIN) at what rev/min*	*MAX. BHP SAE at what rev/min*	*MAX. TORQUE BS (or DIN) at what rev/min lbf ft (kgf m)*	*TORQUE AT GOVERNED SPEED BS (or DIN) lbf ft (kgf m)*	*MINIMUM SPECIFIC FUEL CONSUMPTION at full load lb/bhp h (g/bhp h)*	*DRY WEIGHT with flywheel and starter, but no fan, generator, air filter or clutch lb (kg)*	*MODEL*
5·51 (140)	659 (10·80)	185 2 200	205 —	463 (64) 1 300	435 (60·2)	0·35 (159)	1 790 (810)	OM346
5·51 (140)	659 (10·80)	210 2 200	230 —	535 (74) 1 300	495 (68·5)	0·35 (159)	1 800 (815)	OM346
5·91 (150)	707 (11·58)	240 2 200	265 —	600 (83) 1 300	542 (75)	0·35 (159)	1 920 (870)	OM355
5·51 (140)	532 (8·72)	{172 2 200 192 2 500	190 — 210	427 (59) 1 500	422 (58)	0·35 (159)	1 300 (591)	OM360
5·11 (130)	973 (15·95)	320 2 500	350 —	746 (103) 1 600	710 (98)	0·37 (168)	2 006 (910)	OM403
5·11 (130)	779 (12·76)	256 2 500	280 —	600 (83) 1 600	539 (74·6)	0·37 (168)	1 840 (838)	OM402
4·72 (120)	230 (3·77)	80 2 800	—	174 (24) 1 500	134 (18·6)	0·36 (163)	696 (315)	F4L912
4·72 (120)	345 (5·65)	120 2 800	—	267 (37) 1 500	202 (28·1)	0·35 (159)	882 (400)	F6L912
4·92 (125)	517 (8·48)	170 2 650	—	376 (52) 1 500	338 (46·6)	0·35 (159)	1 235 (560)	F6L413
4·92 (125)	690 (11·31)	232 2 650	—	506 (70) 1 450	460 (63·5)	0·35 (159)	1 586 (720)	F8L413
4·92 (125)	862 (14·13)	305 2 650	—	642 (90) 1 400	568 (78·2)	0·35 (159)	1 810 (820)	F10L413
4·92 (125)	1 035 (16·96)	340 2 650	—	742 (103) 1 500	678 (93·5)	0·35 (159)	2 210 (1 000)	F12L413
4·92 (125)	811 (13·28)	227 —	235 2 200	640 (88·5) 1 550	568 (78·5)	0·36 (163)	2 950 (1 340)	8JV
5·20 (132)	426 (6·99)	160 2 400	—	412 (57) 1 600	350 (48·4)	0·36 (163)	1 420 (648)	CUA
5·91 (150)	673 (11·02)	172 1 900	—	524 (72·5) 1 200	475 (65·7)	0·35 (159)	—	E
5·91 (150)	707 (11·58)	210 2 100	—	600 (82·9) 1 400	525 (72·6)	0·36 (163)	2 070 (940)	E3
5·91 (150)	673 (11·02)	230 1 900	—	708 (98) 1 400	636 (88)	0·35 (159)	2 420 (1 100)	ET

MAKE	*MODEL*	*COOLING* *water or air-cooled*	*CYCLE* *4-stroke or 2-stroke*	*ASPIRATION* *normally asp. supercharged or turbocharged*	*NUMBER OF CYLINDERS*	*LAYOUT* *in-line, vee or horizontally opposed*	*BORE* *in (mm)*
FIAT	803	water	4	N.A.	3	in-line	3·98 (100)
FIAT	804O	water	4	N.A.	4	in-line	3·98 (100)
FIAT	806	water	4	N.A.	6	in-line	3·98 (100)
FIAT	CP3	water	4	N.A.	6	in-line	4·34 (110)
FIAT	203A/61	water	4	N.A.	6	in-line	5·11 (130)
FIAT	203H/61	water	4	N.A.	6	in-line	5·11 (130)
FIAT	8200	water	4	N.A.	6	in-line	4·80 (122)
FIAT	8210	water	4	N.A.	6	in-line	5·20 (137)
FODEN	FD6MkVI	water	2	Scav	6	in-line	3·62 (92)
FODEN	FD6MkVII	water	2	Turb.	6	in-line	3·62 (92)
FORD	2·4	water	4	N.A.	4	in-line	3·69 (93·7
FORD	255 CID	water	4	N.A.	4	in-line	4·22 (107)
FORD	365 CID	water	4	N.A.	6	in-line	4·13 (105)
FORD	380 CID	water	4	N.A.	6	in-line	4·22 (107)
FORD	360 CID Tornado	water	4	Turb.	6	in-line	4·13 (105)

STROKE in (mm)	CAPACITY in³ (l.)	MAX. BHP BS (or DIN) at what rev/min	MAX. BHP SAE at what rev/min	MAX. TORQUE BS (or DIN) at what rev/min lbf ft (kgf m)	TORQUE AT GOV-ERNED SPEED BS (or DIN) lbf ft (kgf m)	MINIMUM SPECIFIC FUEL CONSUMP-TION at full load lb/bhp h (g/bhp h)	DRY WEIGHT with flywheel and starter, but no fan, generator, air filter or clutch lb (kg)	MODEL
4·33 (110)	154 (2·59)	61 3 200	68 —	119 (16·5) 1 800	100 (13·9)	0·40 (182)	485 (220)	803
4·33 (110)	211 (3·45)	81 3 200	90 —	158 (22) 1 800	133 (18·5)	0·38 (172)	618 (280)	804O
4·33 (110)	316 (5·18)	122 3 200	134 —	231 (32) 1 800	201 (27·9)	0·39 (177)	860 (390)	806
5·12 (130)	454 (7·41)	145 2 600	160	355 (49) 1 200	294 (40·8)	0·37 (168)	1 380 (628)	CP3
5·70 (145)	705 (11·55)	194 1 900	210 —	565 (78·3) 900	481 (66·6)	0·37 (168)	2 050 (930)	203A/61
5·70 (145)	705 (11·55)	173 1 900	—	621 (86) 1 100	470 (65·1)	0·38 (172)	2 200 (1 000)	203 H/61
5·52 (140)	598 (9·82)	200 2 500	220 —	506 (70) 1 200	440 (61)	0·37 (168)	1 750 (795)	8200
6·15 (156)	844 (13·80)	260 2 200	285 —	732 (101) 900	650 (90)	0·36 (163)	2 410 (1 095)	8210
4·72 (120)	293 (4·80)	180 2 250	—	440 (61) 1 400	418 (58)	0·41 (187)	1 331 (604)	FD6MkVI
4·72 (120)	293 (4·80)	225 2 200	—	600 (83) 1 300	535 (74)	0·37 (168)	1 379 (625)	FD6MkVII
3·37 (85·6)	144 (2·36)	61 3 600	—	95 (13·6) 2 250	89 (12·4)	0·45 (244)	505 (230)	2·4
4·52 (115)	254 (4·16)	74 2 800	81 —	185 (25·6) 1 600	139 (19·2)	0·39 (177)	741 (337)	255 CID
4·52 (115)	363 (5·95)	103 2 800	115 —	236 (32·6) 1 700	193 (26·7)	0·38 (173)	955 (433)	365 CID
4·52 (115)	380 (6·23)	114 2 800	127 —	273 (37·7) 1 500	214 (29·6)	0·43 (195)	946 (429)	380 CID
4·52 (115)	363 (5·95)	143 2 400	150 —	338 (46·7) 1 800	313 (43·3)	0·37 (168)	1 050 (476)	360 CID Tornado

MAKE	*MODEL*	*COOLING* *water or air-cooled*	*CYCLE* *4-stroke or 2-stroke*	*ASPIRA-TION* *normally asp. super-charged or turbo-charged*	*NUMBER OF CYLIN-DERS*	*LAYOUT* *in-line, vee or horizontally opposed*	*BORE* *in (mm*
GARDNER	4LW20	water	4	N.A.	4	in-line	4·25 (108)
GARDNER	5LW20	water	4	N.A.	5	in-line	4·25 (108)
GARDNER	6LW20	water	4	N.A.	6	in-line	4·25 (108)
GARDNER	6LX	water	4	N.A.	6	in-line	4·75 (121)
GARDNER	6LXB	water	4	N.A.	6	in-line	4·75 (121)
GARDNER	8LXB	water	4	N.A.	8	in-line	4·75 (121)
G.M.C.	3-53	water	2	Scav	3	in-line	3·88 (98)
G.M.C.	4-53	water	2	Scav	4	in-line	3·88 (98)
G.M.C.	3-71	water	2	Scav	3	in-line	4·25 (108)
G.M.C.	4-71	water	2	Scav	4	in-line	4·25 (108)
G.M.C.	6V-53	water	2	Scav	6	66·7° Vee	3·88 (98)
G.M.C.	8V-53	water	2	Scav	8	66·7° Vee	3·88 (98)
G.M.C.	DH478	water	4	N.A.	6	60° Vee	5·13 (130)
G.M.C.	DH637	water	4	N.A.	8	60° Vee	5·13 (130)
G.M.C.	6-71	water	2	Scav	6	in-line	4·25 (108)
G.M.C.	6V-71	water	2	Scav	6	63·5° Vee	4·25 (108)
G.M.C.	8V-71	water	2	Scav	8	63·5° Vee	4·25 (108)

STROKE in (mm)	CAPACITY in^3 (l.)	MAX. BHP BS (or DIN) at what rev/min	MAX. BHP SAE at what rev/min	MAX. TORQUE BS (or DIN) at what rev/min lbf ft (kgf m)	TORQUE AT GOVERNED SPEED BS (or DIN) lbf ft (kgf m)	MINIMUM SPECIFIC FUEL CONSUMPTION at full load lb/bhp hr (g/bhp h)	DRY WEIGHT with flywheel and starter, but no fan, generator, air filter or clutch lb (kg)	MODEL
6·00 (152)	340 (5·58)	80 1 700	—	253 (35) 1 350	247 (33)	0·34 (155)	1 135 (515)	4LW20
6·00 (152)	425 (6·97	100 1 700	—	317 (44) 1 350	310 (43)	0·34 (155)	1 295 (588)	5LW20
6·00 (152)	510 (8·37)	120 1 700	—	381 (52·8) 1 350	372 (51·5)	0·34 (155)	1 485 (675)	6LW20
6·00 (152)	638 (10·45)	150 1 700	—	485 (67) 1 050	464 (64)	0·33 (150)	1 625 (738)	6LX
6·00 (152)	638 (10·45)	180 1 850	—	536 (74·2) 1 050	511 (70·8)	0·33 (150)	1 625 (738)	6LXB
6·00 (152)	851 (13·93)	240 1 850	—	695 (96·1) 1 100	682 (94·3)	0·33 (150)	2 045 (928)	8LXB
4·50 (114)	159 (2·6)	101 2 800	—	216 (29·9) 1 500	189 (27·6)	0·42 (190)	1 000 (454)	3-53
4·50 (114)	212 (3·5)	140 2 800	—	295 (40·8) 1 800	262 (36·2)	0·42 (190)	1 185 (538)	4-53
5·00 (127)	213 (3·5)	106 2 100	—	293 (40·5) 1 400	265 (36·6)	0·45 (204)	1 515 (688)	3-71
5·00 (127)	284 (4·65)	160 2 100	—	432 (60) 1 400	400 (55·5)	0·38 (172)	1 740 (790)	4-71
4·50 (114)	318 (5·2)	195 2 600	—	446 (61·8) 1 500	394 (54·6)	0·40 (182)	1 530 (695)	6V-53
4·50 (114)	425 (6·97)	247 2 500	—	580 (80·4) 1 500	518 (71·8)	0·40 (182)	1 890 (858)	8V-53
3·86 (98)	478 (7·83)	155 2 800	165 —	325 (44·8) 2 000	304 (42)	0·38 (172)	950 (431)	DH478
3·86 (98)	637 (10·43)	195 2 800	210 —	404 (55·9) 1 800	383 (53)	0·38 (172)	1 271 (578)	DH637
5·00 (127)	426 (6·98)	238 2 100	—	649 (90) 1 400	595 (82·5)	0·38 (172)	2 140 (970)	6-71
5·00 (127)	426 (6·98)	238 2 100	—	649 (90) 1 400	595 (82·5)	0·38 (172)	1 950 (884)	6V-71
5·00 (127)	567 (9·3)	318 2 100	—	864 (119) 1 400	794 (110)	0·38 (173)	2 335 (1 060)	8V-71

MAKE	MODEL	COOLING water or air-cooled	CYCLE 4-stroke or 2-stroke	ASPIRATION normally asp. supercharged or turbocharged	NUMBER OF CYLINDERS	LAYOUT in-line, vee or horizontally opposed	BORE in (mm)
G.M.C.	12V-71	water	2	Scav	12	63·5° Vee	4·25 (108)
HANOMAG	D141	water	4	N.A.	4	in-line	3·74 (95)
HANOMAG	D142	water	4	N.A.	4	in-line	3·94 (100)
HANOMAG	D161	water	4	N.A.	6	in-line	3·74 (95)
HANOMAG	D162	water	4	N.A.	6	in-line	3·94 (100)
HENSCHEL	6R1112 (561)	water	4	N.A.	6	in-line	4·52 (115)
HENSCHEL	6R1315 (525)	water	4	N.A.	6	in-line	5·11 (130)
HINO	DM100	water	4	N.A.	6	in-line	3·54 (90)
HINO	DS70	water	4	N.A.	6	in-line	4·13 (105)
HINO	DS5D	water	4	N.A.	6	in-line	4·33 (110)
HINO	EB100	water	4	N.A.	6	in-line	4·53 (115)
HINO	DK10	water	4	N.A.	6	in-line	4·72 (120)
HINO	DK10-2	water	4	Turb.	6	in-line	4·73 (120)
HINO	DA-59	water	4	N.A.	6	in-line	4·72 (120)
HINO	EA100	water	4	N.A.	8	90° Vee	5·51 (140)
HINO	DS120	water	4	N.A.	12	in-line horiz. opp.	4·33 (110)

TROKE n (mm)	CAPACITY in^3 (l.)	MAX. BHP BS (or DIN) at what rev/min	MAX. BHP SAE at what rev/min	MAX. TORQUE BS (or DIN) at what rev/min lbf ft (kgf m)	TORQUE AT GOV- ERNED SPEED BS (or DIN) lbf ft (kgf m)	MINIMUM SPECIFIC FUEL CONSUMP- TION at full load lb/bhp h (g/bhp h)	DRY WEIGHT with flywheel and starter, but no fan, generator, air filter or clutch lb (kg)	MODEL
5·00 (127)	851 (13·9)	475 2 100	—	1 300 (180) 1 400	1 190 (165)	0·38 (173)	3 290 (1 492)	12V-71
3·94 (100)	173 (2·84)	65 3 400	72 —	123 (17) 1 800	98 (13·6)	0·43 (195)	618 (280)	D141
3·94 (100)	190 (3·11)	80 3 400	88 —	144 (20) 1 800	121 (16·8)	0·43 (195)	618 (280)	D142
3·94 (100)	260 (4·26)	100 3 400	110 —	181 (25) 1 800	152 (21)	0·42 (190)	794 (360)	D161
3·94 (100)	288 (4·71)	115 3 400	127 —	217 (30) 1 800	176 (24·3)	0·42 (190)	794 (360)	D162
4·93 (125)	475 (7·80)	180 2 600	196 —	413 (57) 1 400	348 (47·8)	0·35 (157)	1 390 (630)	6R1112 (561)
5·90 (150)	735 (11·94)	242 2 250	263 —	650 (90) 1 250	592 (82)	0·34 (155)	1 800 (815)	6R1315 (525)
4·45 (113)	264 (4·31)	90 3 200	—	181 (25) 2 000	145 (20)	0·42 (190)	882 (400)	DM100
5·31 (135)	428 (7·01)	140 2 500	—	318 (44) 1 600	292 (40·5)	0·42 (190)	1 410 (640)	DS70
5·52 (140)	486 (7·98)	160 2 400	—	372 (51·5) 1 600	347 (48)	0·42 (190)	1 485 (675)	DS5D
5·70 (145)	552 (9·04)	175 2 350	—	423 (58·5) 1 400	383 (53)	0·42 (190)	1 715 (780)	EB100
5·90 (150)	620 (10·18)	195 2 300	—	491 (68) 1 400	405 (56)	0·41 (185)	2 073 (920)	DK10
5·92 (150)	620 (10·18)	230 2 300	—	585 (81) 1 600	520 (72)	0·41 (185)	2 115 (960)	DK10-2
6·30 (160)	663 (10·86)	175 2 000	—	492 (68) 1 400	455 (63)	0·42 (190)	2 315 (1 050)	DA-59
4·33 (110)	828 (13·55)	280 2 600	—	606 (84) 1 800	541 (75)	0·42 (190)	2 538 (1 150)	EA100
5·51 (140)	974 (15·97)	320 2 400	—	744 (103) 1 600	694 (96)	0·43 (195)	3 440 (1 560)	DS120

MAKE	MODEL	COOLING water or air-cooled	CYCLE 4-stroke, or 2-stroke	ASPIRATION normally asp. super-charged or turbo-charged	NUMBER OF CYLINDERS	LAYOUT in-line, vee or horizontally opposed	BORE in (mm
INTERNATIONAL	DV-462	water	4	N.A.	8	90° Vee	4·13 (105)
INTERNATIONAL	DV-550	water	4	N.V.	8	90° Vee	4·50 (114)
INTERNATIONAL	DVT-573	water	4	Turb.	8	90° Vee	4·50 (114)
ISUZU	B110	water	4	N.A.	2	in-line	3·72 (83)
ISUZU	C201	water	4	N.A.	4	in-line	3·72 (83)
ISUZU	C221	water	4	N.A.	4	in-line	3·72 (83)
ISUZU	D400	water	4	N.A.	4	in-line	3·62 (92)
ISUZU	DA220	water	4	N.A.	4	in-line	3·94 (100)
ISUZU	DA120	water	4	N.A.	6	in-line	3·94 (100)
ISUZU	DA640	water	4	N.A.	6	in-line	4·02 (102)
ISUZU	DA640T	water	4	Turb.	6	in-line	4·02 (102)
ISUZU	D920	water	4	N.A.	6	in-line	4·93 (125)
ISUZU	DH100	water	4	N.A.	6	in-line	4·72 (120)
ISUZU	DH100T	water	4	Turb.	6	in-line	4·72 (120)
ISUZU	E110	water	4	N.A.	6	in-line	4·93 (125)
LANCIA	402 000	water	4	N.A.	6	in-line	4·80 (122
LANCIA	520 000	water	4	N.A.	6	in-line	4·80 (122

STROKE in (mm)	*CAPACITY in³ (l.)*	*MAX. BHP BS (or DIN) at what rev/min*	*MAX. BHP SAE at what rev/min*	*MAX. TORQUE BS (or DIN) at what rev/min lbf ft (kgf m)*	*TORQUE AT GOVERNED SPEED BS (or DIN) lbf ft (kgf m)*	*MINIMUM SPECIFIC FUEL CONSUMPTION at full load lb/bhp h (g/bhp h)*	*DRY WEIGHT with flywheel and starter, but no fan, generator, air filter or clutch lb (kg)*	*MODEL*
4·31 (109)	460 (7·55)	—	185 3 200	337 (46·7) 2 100	305 (42·3)	0·40 (182)	1 260 (572)	DV-462
4·31 (109)	550 (9·00)	—	210 3 200	391 (54·1) 2 100	337 (46·6)	0·38 (172)	1 287 (583)	DV-550
4·50 (114)	573 (9·40)	233 2 600	260 —	578 (80) 1 800	470 (65·0)	0·38 (172)	2 220 (1 010)	DVT-573
4·02 (102)	67·3 (1·10)	19 2 600	—	41 (5·7) 2 000	39 (5·4)	0·48 (218)	—	B110
3·62 (92)	121·5 (1·99)	36 2 600	—	79 (10·9) 2 000	76 (10·4)	0·45 (205)	—	C201
4·02 (102)	134·7 (2·21)	48 3 000	—	89 (12·4) 2 000	79 (11)	0·45 (205)	—	C221
3·94 (100)	243 (3·99)	73 2 800	—	152 (21·5) 1 800	142 (19·5)	0·47 (214)	—	D400
5·12 (130)	249 (4·08)	67 2 200	—	174 (24) 1 400	168 (23)	0·45 (205)	—	DA220
5·12 (130)	374 (6·12)	102 2 200	—	267 (37) 1 400	260 (36)	0·42 (190)	—	DA120
5·12 (130)	389 (6·37)	110 2 400	—	274 (37·8) 1 400	252 (35)	0·42 (190)	—	DA640
5·12 (130)	389 (6·37)	130 2 200	—	332 (46) 1 600	325 (45)	0·44 (200)	—	DA640T
4·93 (125)	561 (9·20)	149 2 400	—	358 (49·5) 1 600	342 (47)	0·39 (177)	—	D920
5·91 (150)	621 (10·18)	153 2 000	—	455 (63) 1 400	430 (59·5)	0·41 (187)	—	DH100
5·91 (150)	621 (10·18)	185 2 000	—	563 (78) 1 400	508 (70·5)	0·42 (190)	—	DH100T
5·91 (150)	674 (11·04)	170 2 000	—	494 (68·5) 1 200	469 (65)	0·41 (187)	—	E110
5·31 (135)	578 (9·47)	176 2 300	—	457 (63·3) 900	397 (55)	0·35 (159)	1 400 (635)	402 000
5·91 (150)	643 (10·52)	209 2 200	—	555 (76·8) 900	493 (68·2)	0·37 (168)	1 984 (900)	520 000

MAKE	*MODEL*	*COOLING water or air-cooled*	*CYCLE 4-stroke or 2-stroke*	*ASPIRATION normally asp. super-charged or turbo-charged*	*NUMBER OF CYLINDERS*	*LAYOUT in-line, vee or horizontally opposed*	*BORE in (mm)*
LEYLAND	400	water	4	N.A.	6	in-line	4·22 (107)
LEYLAND	401	water	4	N.A.	6	in-line	4·22 (107)
LEYLAND	410	water	4	Turb.	6	in-line	4·22 (107)
LEYLAND	500	water	4	N.A.	6	in-line	4·65 (118)
LEYLAND	520	water	4	Turb.	6	in-line	4·65 (118)
LEYLAND	600	water	4	N.A.	6	in-line	4·80 (122)
LEYLAND	680	water	4	N.A.	6	in-line	5·00 (127)
LEYLAND	690	water	4	Turb.	6	in-line	5·00 (127)
LEYLAND	L60	water	2	Sup.	6	in-line opp/pist	4·63 (117)
LEYLAND	100 Series	water	4	N.A.	4	in-line	3·25 (82·5)
LEYLAND	120 Series	water	4	N.A.	4	in-line	3·50 (89)
LEYLAND	200 Series	water	4	N.A.	4	in-line	3·96 (100)
LEYLAND	300 series	water	4	N.A.	6	in-line	3·74 (95)
LEYLAND	310 Series	water	4	N.A.	6	in-line	3·94 (100)
M.A.N.	597-02	water	4	N.A.	6	in-line	3·94 (100)
M.A.N.	D0846HM	water	4	N.A.	6	in-line	4·26 (108)
M.A.N.	D0026M	water	4	N.A.	6	in-line	3·94 (100)

TROKE n (mm)	*CAPACITY in^3 (l.)*	*MAX. BHP BS (or DIN) at what rev/min*	*MAX. BHP SAE at what rev/min*	*MAX. TORQUE BS (or DIN) at what rev/min lbf ft (kgf m)*	*TORQUE AT GOV-ERNED SPEED BS (or DIN) lbf ft (kgf m)*	*MINIMUM SPECIFIC FUEL CONSUMP-TION at full load lb/bhp h (g/bhp h)*	*DRY WEIGHT with flywheel and starter, but no fan, generator, air filter or clutch lb (kg)*	*MODEL*
4·75 (121)	399 (6·54)	126 2 400	134 —	315 (43) 1 600	274 (38·0)	0·38 (172)	1 181 (537)	400
4·75 (121)	399 (6·54)	138 2 600	149 —	320 (44·2) 1 600	279 (38·5)	0·38 (172)	—	401
4·75 (121)	399 (6·54)	155 2 600	167 —	375 (52) 1 600	327 (45)	0·36 (163)	1 230 (558)	410
4·92 (125)	500 (8·20)	170 2 600	183 —	375 (52) 1 800	342 (47·3)	0·36 (163)	1 659 (753)	500
4·92 (125)	500 (8·20)	260 2 600	279 —	630 (87·5) 1 700	522 (72)	0·34 (155)	1 680 (764)	520
5·50 (140)	600 (9·80)	166 2 200	180 —	442 (61·2) 1 400	390 (54·0)	0·35 (159)	1 808 (820)	600
5·75 (146)	677 (11·10)	200 2 200	215 —	525 (72·6) 1 400	475 (65·8)	0·38 (172)	2 000 (908)	680
5·75 (146)	677 (11·10)	240 2 200	258 —	650 (90) 1 400	575 (79·6)	0·37 (168)	2 050 (930)	690
5·75 (146)	1 159 (19)	650 2 100	700 —	1 560 (216) 1 400	1 402 (194)	0·38 (172)	4 416 (2 000)	L60
4·00 (102)	133 (2·18)	55 3 500	—	89 (12·3) 2 800	84 (11·63)	0·47 (214)	540 (245)	100 Series
4·00 (102)	154 (2·50)	65 3 500	68 —	116 (16) 2 000	—	0·47 (214)	540 (245)	120 Series
4·72 (120)	231·8 (3·80)	64 2 400	69 —	168 (23·3) 1 750	146 (20·2)	0·39 (177)	728 (330)	200 Series
4·72 (120)	311 (5·10)	83 2 400	89 —	224 (31·1) 1 500	197 (27·3)	0·39 (177)	1 008 (457)	300 Series
4·72 (120)	345·3 (5·66)	105 2 400	110 —	255 (35·4) 1 750	230 (31·9)	0·39 (177)	1 008 (457)	310 Series
4·41 (112)	322 (5·28)	126 2 900	139 —	253 (35) 1 500	239 (33)	0·37 (168)	1 188 —	597-02
5·20 (132)	442 (7·25)	160 2 500	172 —	355 (49) 1 600	352 (48·5)	0·36 (163)	1 362 (620)	D0846HM
4·92 (125)	358 (5·89)	115 2 500	127 —	275 (38) 1 400	232 (32)	0·35 (159)	1 230 (558)	D0026M

MAKE	*MODEL*	*COOLING water or air-cooled*	*CYCLE 4-stroke or 2-stroke*	*ASPIRATION normally asp. super-charged or turbo-charged*	*NUMBER OF CYLINDERS*	*LAYOUT in-line, vee or horizontally opposed*	*BORE in (mm*
M.A.N.	D2146M	water	4	N.A.	6	in-line	4·77 (121)
M.A.N.	D2356HM	water	4	N.A.	6	in-line	4·84 (123)
M.A.N.	D2156HM	water	4	N.A.	6	in-line	4·77 (121)
M.A.N.	D2156MT	water	4	Turb.	6	in-line	4·84 (123)
M.A.N.	D2858M	water	4	N.A.	8	90° Vee	5·04 (128)
NISSAN	SD20	water	4	N.A.	4	in-line	3·27 (83)
NISSAN	SD22	water	4	N.A.	4	in-line	3·27 (83)
NISSAN	SD33	water	4	N.A.	6	in-line	3·27 (83)
NISSAN	UD33N	water	2	Scav Sup.	3	in-line	4·34 (110)
NISSAN	UD43N	water	2	Scav Sup.	4	in-line	4·34 (110)
NISSAN	UD50N	water	2	Scav Sup.	5	in-line	4·34 (110)
NISSAN	UD63	water	2	Scav Sup.	6	in-line	4·34 (110)
NISSAN	UDV81	water	2	Scav Sup.	8	90° Vee	4·34 (110
Ö.A.F.	9.5	water	4	N.A.	6	in-line	4·72 (120
Ö.A.F.	10.3	water	4	N.A.	6	in-line	4·76 (121

STROKE in (mm)	*CAPACITY in³ (l.)*	*MAX. BHP BS (or DIN) at what rev/min*	*MAX. BHP SAE at what rev/min*	*MAX. TORQUE BS (or DIN) at what rev/min lbf ft (kgf m)*	*TORQUE AT GOV-ERNED SPEED BS (or DIN) lbf ft (kgf m)*	*MINIMUM SPECIFIC FUEL CONSUMP-TION at full load lb/bhp h (g/bhp h)*	*DRY WEIGHT with flywheel and starter, but no fan, generator, air filter or clutch lb (kg)*	*MODEL*
5·51 (140)	589 (9·66)	186 2 200	205 —	477 (66) 1 400	441 (61)	0·36 (163)	1 629 (740)	D2146M
5·91 (150)	653 (10·69)	230 2 200	254 —	594 (82) 1 600	539 (74·5)	0·36 (163)	1 650 (750)	D2356HM
5·91 (150)	635 (10·34)	212 2 200	236 —	550 (76) 1 400	506 (70)	0·34 (155)	1 650 (750)	D2156HM
5·91 (150)	651 (10·69)	256 2 200	270 —	662 (91·5) 1 400	640 (88·5)	0·35 (159)	1 693 (770)	D2156MT
5·91 (150)	973 (15·45)	304 2 200	320 —	825 (114) 1 600	760 (105)	0·36 (163)	2 340 (1 062)	D2858M
3·62 (92)	121 (1·99)	60 4 000	—	94 (13) 1 800	—	0·43 (195)	411 (187)	SD20
3·94 (100)	132 (2·16)	65 4 000	—	105 (14·5) 1 800	—	0·43 (195)	418 (190)	SD22
3·94 (100)	198 (3·24)	98 4 000	—	159 (22) 1 800	—	0·43 (195)	591 (268)	SD33
5·12 (130)	226 (3·70)	130 2 400	—	328 (45·5) 1 400	—	0·41 (187)	1 232 (560)	UD33N
5·12 (130)	301 (4·94)	175 2 400	—	456 (63) 1 400	—	0·41 (187)	1 441 (655)	UD43N
5·12 (130)	377 (6·18)	215 2 400	—	542 (75) 1 400	—	0·41 (187)	1 629 (740)	UD50N
5·12 (130)	452 (7·41)	240 2 200	—	665 (92) 1 400	—	0·42 (190)	2 107 (960)	UD63
5·12 (130)	601 (9·88)	330 2 200	—	884 (122) 1 400	—	0·42 (190)	2 640 (1 200)	UDV81
5·51 (140)	580 (9·50)	180 2 100	—	478 (66) 1 400	441 (61)	0·40 (182)	1 870 (850)	9.5
5·90 (150)	6·31 (10·35)	215 2 200	—	550 (76) 1 500	513 (71)	0·37 (168)	2 080 (950)	10.3

MAKE	*MODEL*	*COOLING* *water or air-cooled*	*CYCLE* *4-stroke or 2-stroke*	*ASPIRATION* *normally asp. supercharged or turbocharged*	*NUMBER OF CYLINDERS*	*LAYOUT* *in-line, vee or horizontally opposed*	*BORE* *in (mm)*
PEGASO	9040	water	4	N.A.	4	in-line	4·21 (107)
PEGASO	9042	water	4	N.A.	4	in-line	4·21 (107)
PEGASO	9100	water	4	N.A.	6	in-line	4·65 (118)
PEGASO	9105	water	4	N.A.	6	in-line	4·72 (120)
PEGASO	9109	water	4	Turb.	6	in-line	4·72 (120)
PEGASO	9156	water	4	Turb.	6	in-line	5·12 (130)
PERKINS	4.108	water	4	N.A.	4	in-line	3·13 (79)
PERKINS	4.154	water	4	N.A.	4	in-line	3·50 (89)
PERKINS	4.203	water	4	N.A.	4	in-line	3·60 (92)
PERKINS	4.236	water	4	N.A.	4	in-line	3·88 (98)
PERKINS	6.354	water	4	N.A.	6	in-line	3·88 (98)
PERKINS	T6.3543	water	4	Turb.	6	in-line	3·88 (98)
PERKINS	V8-510	water	4	N.A.	8	90° Vee	4·25 (108)
PERKINS	V8.605	water	4	N.A.	8	90° Vee	4·50 (114)
ROLLS-ROYCE	Eagle 205	water	4	N.A.	6	in-line	5·13 (130)
ROLLS-ROYCE	Eagle 220	water	4	N.A.	6	in-line	5·13 (130)
ROLLS-ROYCE	Eagle 265	water	4	Turb.	6	in-line	5·13 (130)

STROKE *in (mm)*	*CAPACITY* *in*³ *(l.)*	*MAX. BHP BS (or DIN) at what rev/min*	*MAX. BHP SAE at what rev/min*	*MAX. TORQUE BS (or DIN) at what rev/min lbf ft (kgf m)*	*TORQUE AT GOV-ERNED SPEED BS (or DIN) lbf ft (kgf m)*	*MINIMUM SPECIFIC FUEL CONSUMP-TION at full load lb/bhp h (g/bhp h)*	*DRY WEIGHT with flywheel and starter, but no fan, generator, air filter or clutch lb (kg)*	*MODEL*
4·76 (121)	267 (4·37)	—	95 2 200	221 (30·5) 1 400	—	0·36 (163)	900 (408)	9040
4·76 (121)	267 (4·37)	—	90 2 400	221 (30·5) 1 400	197 (22·2)	0·36 (163)	915 (415)	9042
6·10 (155)	621 (10·17)	—	185 2 100	474 (65·5) 1 300	464 (64)	0·36 (163)	1 444 (655)	9100
6·10 (155)	646 (10·52)	—	200 2 000	542 (75) 1 100	528 (73)	0·36 (163)	1 676 (760)	9105
6·10 (155)	646 (10·52)	—	260 2 000	722 (100) 1 400	686 (95)	0·37 (166)	2 150 (975)	9109
5·90 (150)	729 (11·95)	—	352 2 200	932 (129) 1 400	840 (116)	0·35 (158)	2 308 (1 027)	9156
3·50 (89)	108 (1·76)	50 4 000	—	80 (11·0) 2 000	65 (9·0)	0·42 (190)	390 (177)	4.108
4·00 (102)	154 (2·52)	68 3 600	—	112 (15·5) 2 000	101 (13·9)	0·40 (182)	510 (232)	4.154
5·00 (127)	203 (3·33)	59 2 600	—	150 (20·7) 1 350	118 (16·3)	0·42 (190)	570 (258)	4.203
5·00 (127)	236 (3·86)	78 2 800	—	193 (26·7) 1 400	145 (20)	0·36 (163)	670 (304)	4.236
5·00 (127)	354 (5·80)	118 2 800	—	262 (36·2) 1 250	210 (30·2)	0·37 (168)	925 (420)	6.354
5·00 (127)	354 (5·80)	143 2 600	—	358 (49·5) 1 600	285 (39·4)	0·37 (168)	1 060 (482)	T6.3543
4·50 (115)	510 (8·36)	167 2 800	—	380 (52·5) 1 600	309 (43)	0·38 (172)	1 440 (655)	V8-510
4·75 (121)	605 (9·9)	205 2 600	—	460 (62) 1 650	415 (57)	0·38 (172)	1 580 (718)	V8.605
6·00 (152)	743 (12·17)	205 2 100	—	559 (77) 1 400	537 (74)	0·37 (168)	2 410 (1 092)	Eagle 205
6·00 (152)	743 (12·17)	220 2 100	—	605 (83) 1 400	576 (79·5)	0·38 (172)	2 410 (1 092)	Eagle 220
6·00 (152)	743 (12·17)	265 2 100	—	755 (104) 1 400	694 (96)	0·36 (163)	2 562 (1 161)	Eagle 265

MAKE	*MODEL*	*COOLING* *water or air-cooled*	*CYCLE* *4-stroke or 2-stroke*	*ASPIRATION* *normally asp. super-charged or turbo-charged*	*NUMBER OF CYLINDERS*	*LAYOUT* *in-line, vee or horizontally opposed*	*BORE* *in (mm*
ROLLS-ROYCE	Eagle 305	water	4	Turb.	6	in-line	5·13 (130)
ROVER	2·5 l.	water	4	N.A.	4	in-line	3·56 (90·50
SAURER	CT4D	water	4	N.A.	6	in-line	4·53 (115)
SAURER	CT5D	water	4	N.A.	6	in-line	4·53 (115)
SAURER	D1K	water	4	N.A.	6	in-line	5·12 (130)
SAURER	D1KL	water	4	Sup.	6	in-line	5·05 (128)
SAURER	D1KT	water	4	Turb.	6	in-line	5·05 (128)
SCANIA	D5	water	4	N.A.	4	in-line	4·53 (115)
SCANIA	DS5	water	4	Turb.	4	in-line	4·53 (115)
SCANIA	D8	water	4	N.A.	6	in-line	4·53 (115)
SCANIA	DS8	water	4	Turb.	6	in-line	4·53 (115)
SCANIA	D11	water	4	N.A.	6	in-line	5·00 (127)
SCANIA	DS11	water	4	Turb.	6	in-line	5·00 (127)
SCANIA	DS14	water	4	Turb.	8	90° Vee	5·00 (127)
STEYR	WD609	water	4	N.A.	6	in-line	4·13 (105)
STEYR	WD609.60	water	4	Turb.	6	in-line	4·13 (105)
STEYR	WD614.01	water	4	N.A.	6	in-line	4·72 (120)

STROKE in (mm)	CAPACITY in^3 (l.)	MAX. BHP BS (or DIN) at what rev/min	MAX. BHP SAE at what rev/min	MAX. TORQUE BS (or DIN) at what rev/min lbf ft (kgf m)	TORQUE AT GOV-ERNED SPEED BS (or DIN) lbf ft (kgf m)	MINIMUM SPECIFIC FUEL CONSUMP-TION at full load lb/bhp h (gm/bhp h)	DRY WEIGHT with flywheel and starter, but no fan, generator, air filter or clutch lb (kg)	MODEL
6·00 (152)	743 (12·17)	305 2 100	—	855 (118) 1 400	800 (110)	0·35 (159)	2 562 (1 161)	Eagle 305
3·50 (89·00)	140 (2·29)	67 4 000	—	105 (14·5) 1 800	81·5 (11·3)	0·51 (231)	533 (242)	2·5 l.
4·72 (120)	416 (7·48)	130 2 200	—	333 (46) 1 300	326 (45)	—	—	CT4D
5·51 (140)	532 (8·73)	160 2 200	—	434 (60) 1 300	383 (53)	—	—	CT5D
5·91 (150)	730 (11·95)	230 2 200	—	550 (76) 1 300	575 (79·5)	0·35 (160)	—	D1K
5·91 (150)	708 (11·60)	270 2 000	—	724 (100) 1 200	689 (95)	0·37 (168)	—	D1KL
5·91 (150)	708 (11·6)	310 2 200	—	822 (114) 1 450	742 (101·3)	0·35 (157)	—	D1KT
4·92 (125)	318 (5·20)	102 2 400	107 —	275 (38) 1 200	224 (31)	0·38 (172)	1 145 (520)	D5
4·92 (125)	318 (5·20)	130 2 400	135 —	318 (44) 1 500	278 (38·5)	0·38 (172)	1 200 (545)	DS5
4·92 (125)	475 (7·79)	155 2 400	167 —	405 (56) 1 200	355 (49)	0·37 (168)	1 510 (685)	D8
4·92 (125)	475 (7·79)	192 2 400	204 —	499 (69) 1 500	440 (61)	0·37 (168)	1 671 (760)	DS8
5·71 (145)	674 (11·02)	195 2 200	215 —	572 (79) 1 200	455 (63)	0·38 (172)	1 905 (865)	D11
5·71 (145)	674 (11·02)	275 2 200	285 —	780 (108) 1 400	651 (90)	0·36 (163)	1 955 (890)	DS11
5·51 (140)	868 (14·20)	350 2 300	365 —	957 (127) 1 450	837 (115)	0·35 (159)	2 553 (1 160)	DS14
4·53 (115)	366 (5·98)	150 2 800	—	313 (43·5) 1 600	295 (41)	—	—	WD609
4·53 (115)	366 (5·98)	150 2 800	162 —	315 (43·5) 1 700	282 (39)	0·38 (172)	—	WD609.60
4·72 (120)	496 (8·14)	180 2 800	198 —	375 (52·0) 1 400	338 (46·8)	0·37 (168)	1 675 (760)	WD614.01

MAKE	*MODEL*	*COOLING* *water or air-cooled*	*CYCLE* *4-stroke or 2-stroke*	*ASPIRATION* *normally asp. supercharged or turbocharged*	*NUMBER OF CYLINDERS*	*LAYOUT* *in-line, vee or horizontally opposed*	*BORE* *in (mm)*
STEYR	WD614.60	water	4	Turb.	6	in-line	4·72 (120)
STEYR	WD815.01	water	4	N.A.	8	90° Vee	4·96 (126)
STEYR	WD815.60	water	4	Turb	8	90° Vee	4·96 (126)
TATRA	T924	air	4	N.A.	4	in-line	4·72 (120)
TATRA	T928	air	4	N.A.	8	75° Vee	4·72 (120)
TATRA	T2-138	water	4	N.A.	8	75° Vee	4·72 (120)
TATRA	T930-3	air	4	N.A.	12	75° Vee	4·72 (120)
UNIC	32SD	water	4	N.A.	4	in-line	4·69 (119)
UNIC	32S	water	4	N.A.	4	in-line	4·69 (119)
UNIC	42SB	water	4	N.A.	6	in-line	4·69 (119)
UNIC	V85S	water	4	N.A.	8	90° Vee	5·32 (135)
UNIC	V62S	water	4	N.A.	8	90° Vee	4·69 (119)
VOLVO	TD50	water	4	Turb.	6	in-line	3·75 (95)
VOLVO	TD70B	water	4	Turb.	6	in-line	4·13 (105)
VOLVO	TD100	water	4	Turb.	6	in-line	4·76 (121)
VOLVO	TD120	water	4	Turb.	6	in-line	5·13 (130)

STROKE in (mm)	*CAPACITY in³ (l.)*	*MAX. BHP BS (or DIN) at what rev/min*	*MAX. BHP SAE at what rev/min*	*MAX. TORQUE BS (or DIN) at what rev/min lbf ft (kgf m)*	*TORQUE AT GOV-ERNED SPEED BS (or DIN) lbf ft (kgf m)*	*MINIMUM SPECIFIC FUEL CONSUMP-TION at full load lb/bhp h (g/bhp h)*	*DRY WEIGHT with flywheel and starter, but no fan, generator, air filter or clutch lb (kg)*	*MODEL*
4·72 (120)	496 (8·14)	230 2 600	253 —	505 (70) 1 600	466 (64·5)	0·37 (168)	1 765 (802)	WD614.60
4·72 (120)	734 (12·00)	260 2 600	—	586 (81) 1 500	550 (76)	0·38 (172)	2 098 (955)	WD815·01
4·72 (120)	734 (12·00)	320 2 600	—	723 (100) 1 600	678 (94)	0·38 (172)	2 100 (980)	WD815.60
5·12 (130)	358 (5·88)	70 1 800	—	224 (31) 1 200	214 (29·5)	0·38 (172)	848 (385)	T924
5·12 (130)	718 (11·76)	192 2 000	—	550 (76) 1 200	528 (73)	0·39 (177)	1 434 (650)	T928
5·51 (140)	772 (12·66)	200 2 000	—	593 (82) 1 200	—	0·39 (177)	1 926 (875)	T2-138
5·12 (130)	1 077 (17·64)	250 2 000	—	709 (98) 1 200	688 (95)	0·38 (172)	2 280 (1 035)	T930-3
3·98 (101)	274 (4·49)	100 2 600	110 —	231 (32) 1 400	203 (28)	0·37 (168)	—	32SD
4·76 (121)	328 (5·38)	125 2 600	135 —	304 (42) 1 400	267 (37)	0·36 (163)	—	32S
4·76 (121)	493 (8·07)	180 2 600	200 —	404 (56) 1 400	365 (50·5)	0·36 (163)	—	42SB
5·12 (130)	910 (14·88)	310 2 400	340 —	712 (98) 1 300	680 (94)	0·36 (163)	2 178 (990)	V85S
4·76 (121)	657 (10·77)	248 2 600	270 —	536 (74) 1 400	487 (67·3)	0·36 (163)	1 728 (786)	V62S
4·72 (120)	311 (5·10)	152 2 800	165 —	326 (45) 1 700	298 (41)	0·35 (159)	1 246 (565)	TD50
5·12 (130)	409 (6·70)	207 2 400	210 —	506 (70) 1 400	474 (65·5)	0·35 (159)	1 432 (650)	TD70B
5·51 (140)	586 (9·60)	260 2 200	270 —	694 (96) 1 400	650 (90)	0·35 (159)	2 020 (915)	TD100
5·91 (150)	732 (12·00)	335 2 200	330 —	927 (128) 1 300	825 (114)	0·35 (159)	2 284 (1 036)	TD120

MAKE	*MODEL*	*COOLING water or air-cooled*	*CYCLE 4-stroke or 2-stroke*	*ASPIRATION normally asp. super-charged or turbo-charged*	*NUMBER OF CYLINDERS*	*LAYOUT in-line, vee or horizontally opposed*	*BORE in (mm)*
YAROSLAVL	YMZ-236	water	4	N.A.	6	90° Vee	5·12 (130)
YAROSLAVL	YMZ-238	water	4	N.A.	8	90° Vee	5·12 (130)
YAROSLAVL	YMZ-238N	water	4	Turb.	8	90° Vee	5·12 (130)
YAROSLAVL	YMZ-238NB	water	4	Turb.	8	90° Vee	5·12 (130)

STROKE in (mm)	*CAPACITY in³ (l.)*	*MAX. BHP BS (or DIN) at what rev/min*	*MAX. BHP SAE at what rev/min*	*MAX. TORQUE BS (or DIN) at what rev/min lbf ft (kgf m)*	*TORQUE AT GOV-ERNED SPEED BS (or DIN) lbf ft (kgf m)*	*MINIMUM SPECIFIC FUEL CONSUMP-TION at full load lb/bhp h (g/bhp h)*	*DRY WEIGHT with flywheel and starter, but no fan, generator, air filter or clutch lb (kg)*	*MODEL*
5·51 (140)	680 (11·15)	180 2 100	—	493 (68) 1 500	451 (62)	0·37 (168)	1 870 (850)	YMZ-236
5·51 (140)	907 (14·86)	240 2 100	—	650 (90) 1 500	602 (83)	0·37 (168)	2 290 (1 040)	YMZ-238
5·51 (140)	907 (14·86)	300 2 100	—	795 (110) 1 500	752 (104)	0·37 (168)	2 490 (1 130)	YMZ-238N
5·51 (140)	907 (14·86)	215 1 700	—	695 (96) 1 100	667 (92)	0·37 (168)	2 490 (1 130)	YMZ-238NB

Appendix 2: METRIC CONVERSION TABLE

Bore or Stroke		*Swept volume*		*Popular description*
mm	*cm*^{3}	*cm*^{3}	*in*^{3}	
73	—	1 750	110	1·7 (or 1·8) litre
76	3	2 246	—	2¼ litre
81	—	2 446	150	2½ litre
82	3¼	3 150	200	3 litre
87	—	3 830	—	
89	3½	4 090	250	4 litre
91	—	4 411	—	
95	3¾	4 880	—	
98	—	4 891	300	5 litre or '300'
102	4	5 401	—	
108	4¼	5 760	350	'350'
110	—	6 502	400	
113·3	—	6 750	—	
114·3	4½	7 350	450	
117·4	—	7 883	—	7·8 litre
120·6	—	9 636	—	9·6 litre
122	4¾	9 783	600	9·8 litre or '600'
125	—	10 618	650	10 litre or '650'
127	—	11 100	—	
130	5	11 300	680	'680'
146	5¾	13 743	700	
152·4	6	15 000	900	'900'

Specification of cylinder dimensions in either metric or English notation varies from maker to maker and from country to country. For precise conversion reference to standard tables must be made, but for quick comparison this list of approximate equivalents covering the size range common to automotive diesels will prove useful.

Index